Encyclical on Climate Change & Inequality

Pope Francis

Encyclical on Climate Change & Inequality

On Care for Our Common Home

Introduction by Naomi Oreskes

MELVILLE HOUSE
BROOKLYN · LONDON

Encyclical
on Climate Change
& Inequality

Copyright © 2015 by Libreria Editrice Vaticana
This edition copyright © 2015 by Melville House Publishing LLC

First Melville House Printing: July 2015

Melville House Publishing   8 Blackstock Mews
        46 John Street   and   Islington
    Brooklyn, NY 11201   London N4 2BT

mhpbooks.com   facebook.com/mhpbooks   @melvillehouse

Library of Congress Cataloging-in-Publication Data

Catholic Church. Pope (2013– : Francis)
    [Laudato si'. English]
    Encyclical on climate change and inequality : on care for our
common home / Pope Francis ; introduction by Naomi Oreskes.
        pages   cm
    Includes bibliographical references.
    ISBN 978-1-61219-528-5 (pbk.)
    ISBN 978-1-61219-529-2 (ebook)
    1. Human ecology—Religious aspects—Catholic Church.
2. Ecotheology.   3. Climatic changes—Effect of human beings
on.   4. Christian sociology—Catholic Church.   5. Catholic
Church—Doctrines.   I. Title.

BX1795.H82C38313 2015
261.8'8—dc23

                                                    2015026685

Printed in the United States of America
1 3 5 7 9 10 8 6 4 2

Contents

Introduction

by Naomi Oreskes

Historians looking back often recognize turning points, but ordinary people living through them rarely do. Sometimes, however, a book catalyzes thought into action. *Uncle Tom's Cabin* did this, and so did *Silent Spring*: they called attention to facts that had long been known but upon which people had failed to act. Like those works, Pope Francis's *Encyclical* is a call to action that insists we embrace the moral dimensions of problems that have heretofore been viewed primarily as scientific, technological, and economic.

While it has been billed as an encyclical on the environment, the letter covers virtually every important topic in contemporary life. Climate change, deforestation, and the need for clean, safe drinking water are, of course, included here, as is population (and abortion), but also various problems of science and technology—including public transportation, urban planning and architecture, social media, genetic modification of crops, embryonic stem cell research—and law, economy, and governance—including the problems of deregulated markets, corruption, and weak governance.

Two lines of thought particularly stand out. The first is an affirmation of our interconnectedness and mutual responsibility toward one another, as well as toward our common Earthly home. The second is a denunciation of the aspects of modern life that have led to our current predicament. The essence of the critique is that our situation is not an accident—it is the consequence of the way we think and act: we deny the moral dimensions of our decisions and conflate progress with activity. We cannot continue to think and act this way—to disregard both nature and justice—and expect to flourish. It is not only not moral, it is not even rational.

INTERCONNECTEDNESS AND THE COMMON GOOD

The wide-ranging character of the encyclical is consistent with its central, anti-reductionist argument, which is, quite simply, that everyone and everything is related because it is all part of Creation. In this, Pope Francis finds the fundamental basis for human dignity and for our obligation to care for one another, the planet, and the diverse creatures on it.

The early portion of the encyclical is heavy with biblical reference, reminding us that the idea of interconnectedness is not new and, perhaps, not discretionary. St. Francis, the twelfth-century founder of the Franciscan order, patron saint of animals and the environment, and, of course, the pope's namesake, was renowned for his reverence for nature. "Whenever he would gaze at the sun, the moon or the smallest of animals, he [would] burst into song, drawing all other creatures into his praise. He communed with all creation . . . [and] felt called to care for all that exists."

Nor was Francis idiosyncratic. "The Psalms frequently exhort us," Pope Francis reminds us, "to praise God the Creator, 'who spread out the earth on the waters' (*Ps* 136:6)." Leviticus not only instructs us regarding the familiar weekly Sabbath, on which—like the Lord—we should rest, but also the seven-year sabbatical for the land "when sowing was forbidden and one reaped only what was necessary to live on and to feed one's household (*Lev* 25:1–4)," as well as the Jubilee, celebrated after seven weeks of years (i.e., seven x seven, or forty-nine, years) as a year of general forgiveness and "liberty throughout the land for all its inhabitants." These various instructions all remind us to respect "the rhythms inscribed in nature by the hand of the Creator."

While the pope is critical of instrumentalism (more on that later) and underscores the need "for openness to categories which transcend the language of mathematics and biology, and take us to the heart of what it is to be human," he also invokes a theme that has been common in the history of science: that nature is "a magnificent book in which God speaks to us and grants us a glimpse of his infinite beauty and goodness. 'Through the greatness and the beauty of creatures one comes to know by analogy their maker' (*Wis* 13:5); indeed, 'his eternal power and divinity have been made known through his works since the creation of the world' (*Rom* 1:20)." In the eighteenth and nineteenth centuries, it was a commonplace among naturalists that scientific investigations were a means to come closer to God—to know him through his work as well as his word. As Handel put it in his famous oratorio, "The heavens are telling the glory of god, the wonder of his work displays the firmament."

The core of the argument is that *because* human dignity finds its roots in our common Creation, caring for our fellow citizen and caring for our environment *are the same thing.* It is not a question of people versus the environment and choosing which is more important. It is a question of abandoning the notion of "versus" altogether. Respect for creation and respect for human dignity are two aspects of the same idea.

Pope Francis unreservedly rejects the claim that man has the right to use nature as he wills because of the biblical instruction to "have dominion over nature." He characterizes this as an "inadequate presentation of Christian anthropology [that] gave rise to a wrong understanding of the relationship between human beings and the world." He continues: "Often, what was handed on was a Promethean vision of mastery over the world, which gave the impression that the protection of nature was something that only the faint-hearted cared about. Instead, our 'dominion' over the universe should be understood more properly in the sense of responsible stewardship."

The pope unequivocally rejects the accusation that Western theology sanctions—and therefore is to *blame* for—environmental despoliation. Because this view is so widespread—both among Christians (who use it as justification for "putting people first" or "caring about people more") and by non-Christians (in a critique of such attitudes)—the pope's rebuttal is worth citing at length:

We are not God. The earth was here before us and it
has been given to us. This allows us to respond to the
charge that Judaeo-Christian thinking, on the basis
of the Genesis account which grants man "dominion" over the earth (cf. *Gen* 1:28), has encouraged the
unbridled exploitation of nature by painting him as
domineering and destructive by nature. This is not a
correct interpretation of the Bible as understood by the
Church. Although it is true that we Christians have at
times incorrectly interpreted the Scriptures, nowadays
we must forcefully reject the notion that our being created in God's image and given dominion over the earth
justifies absolute domination over other creatures. The
biblical texts . . . tell us to "till and keep" the garden of
the world (cf. *Gen* 2:15). "Tilling" refers to cultivating,
ploughing or working, while "keeping" means caring,
protecting, overseeing and preserving. This implies a
relationship of mutual responsibility between human
beings and nature. Each community can take from the
bounty of the earth whatever it needs for subsistence,
but it also has the duty to protect the earth and to ensure its fruitfulness for coming generations.

In numerous passages supported with Biblical reference,
the pope insists that Scripture offers no warrant for misuse or
destruction of nature. In each case, he returns to the theme of
interconnection, stressing that "human life is grounded in three
fundamental and closely intertwined relationships: with God,
with our neighbour and with the earth itself." The latter is not
just the physical Earth, but all the living creatures on it, as we

are reminded that the Sabbath is not only for us, but also so "that your ox and your donkey may have rest" (*Ex* 23:12). Clearly "the Bible has no place for a tyrannical anthropocentrism unconcerned for other creatures."

To the extent that God has given man a special place, it is the place of "cooperator with God in the work of creation." But that is a responsibility as much as a right, "a splendid universal communion" that entails an obligation of care: "We do not understand our superiority as a reason for personal glory or irresponsible dominion, but rather as a different capacity which, in its turn, entails a serious responsibility stemming from our faith."

If this encyclical were only intended to reach the faithful, the pope might have ended his discussion there. But his intent to reach further becomes clear in the second part of his argument: that current attitudes and behaviors are not only wrong in a moral sense, they are also wrong in a practical sense. They have not *worked*. They have given us a world that is broken and unjust: where the wealthy worry about obesity while the extreme poor have next to nothing, and the whole planet (in what is destined to be the most quoted phrase of the letter) "is beginning to look more and more like an immense pile of filth." This is no accident, for when "nature is viewed solely as a source of profit and gain, this has serious consequences for society."

THE TECHNOCRATIC PARADIGM

It is hardly a surprise that the pope, head of one of the world's most conservative institutions, finds much to dislike in the modern world. What is a surprise—and a serious challenge to our global political and business leadership—is his attention to the set of mentalities that he variously calls the myths of modernity, the myth of progress, and the technocratic paradigm.

Perhaps the most radical part of the letter (and the part that has already proved disturbing to some readers) is its powerful critique of our "models of production and consumption." The pope addresses head-on our prevailing economic practices and the modes of thought that insist—despite considerable evidence to the contrary—that we just need to let markets do their "magic."

While the word "capitalism" does not appear in the letter, the word "market" (or its variants) appears nineteen times, usually in a critical context. The pope is not advocating communism, but he is asking us to acknowledge that we live in a world where the ideology of the marketplace is so dominant that most of us can scarcely imagine an alternative, and where those who try are dismissed as unrealistic, irrational, naïve, faint-hearted, sentimental, romantic, out of step, or (if in America) communists.[i] He is asking us to reexamine the creed of "individualism, unlimited progress, competition, consumerism, the unregulated market."

Consistent with the complex connotations of the word "paradigm," the technocratic paradigm is not a single idea but a conceptual matrix that begins with an overreliance on scientific modes of thought, moves quickly through instrumentalism, and lodges in a worldview that denies both the idea of limits and the visible consequences of ignoring them:

The basic problem . . . is the way that humanity has taken up technology and its development *according to an undifferentiated and one-dimensional paradigm.* This paradigm exalts the concept of a subject who, using logical and rational procedures, progressively approaches and gains control over an external object . . . Men and women have constantly intervened in nature, but for a long time this meant being in tune with and respecting the possibilities offered by the things themselves. It was a matter of receiving what nature itself allowed, as if from its own hand. Now, by contrast, we are the ones to lay our hands on things, attempting to extract everything possible from them while frequently ignoring or forgetting the reality in front of us . . . This has made it easy to accept the idea of infinite or unlimited growth, which proves so attractive to economists, financiers and experts in technology. It is based on the lie that there is an infinite supply of the earth's goods, and this leads to the planet being squeezed dry beyond every limit. It is the false notion that "an infinite quantity of energy and resources are available, that it is possible to renew them quickly, and that the negative effects of the exploitation of the natural order can be easily absorbed."

While the pope labels the paradigm technocratic, the criticism is not of technology per se, but of an uncritical techno-fideism—a "blind confidence in technical solutions"—and of the market logic that harnesses it with a single-minded focus on profit, without thinking about the actual goals of human activity.

The economy accepts every advance in technology with a view to profit, without concern for its potentially negative impact on human beings. Finance overwhelms the real economy . . . Some circles maintain that current economics and technology will solve all environmental problems, and . . . that the problems of global hunger and poverty will be resolved simply by market growth . . . Yet by itself the market cannot guarantee integral human development and social inclusion. At the same time, we have "a sort of 'superdevelopment' of a wasteful and consumerist kind which forms an unacceptable contrast with the ongoing situations of dehumanizing deprivation," while we are all too slow in developing economic institutions and social initiatives which can give the poor regular access to basic resources. We fail to see the deepest roots of our present failures, which have to do with the direction, goals, meaning and social implications of technological and economic growth.

The problematic assumptions are thus three-fold: (1) that everything is here for our use, and whoever can find and market that use is warranted in doing so; (2) that the system that created these problems will somehow also solve them; and (3) that technology, enabled by science and fostered by the profit motive and consumerism, is the foundation of progress, prosperity, freedom, and even happiness.

The pope endorses "technological innovations which can bring about an improvement in the quality of life," and he makes clear that there is a place for properly regulated markets—indeed, he stresses that the poor need more access to them. What he rejects is the logic that sees the marketplace as the solution to all problems, that prioritizes profits to the exclusion of other considerations, and that privileges individual desire over the common good.

The pope also rejects the equating of technological advance with human advance "as if reality, goodness and truth automatically flow from technological and economic power as such," as well as the myth of technological neutrality. In a sentence that might have been written by a historian of science and technology, he insists that we must "accept that technological products are not neutral, for they create a framework which ends up conditioning lifestyles and shaping social possibilities along the lines dictated by the interests of certain powerful groups. Decisions which may seem purely instrumental are in reality decisions about the kind of society we want to build."[ii]

Economists and other "realists" insist that their worldview is nonideological: that capitalist liberal democracy is the inevitable end point of human development, and that attempts to direct the economic system towards more equitable outcomes at best gum up the works, and at worst are fatally counterproductive. The failures of communism are taken as total refutation of any attempted intervention in the marketplace, any attempt to guide technological development towards more humane ends. But theirs is the ideology of no ideology.[iii] Thus it is significant that the pope's critique is based not only on theological foundations, but on empirical ones as well. It is based on the simple fact that the system as it currently operates has failed in three important ways.

The first is a failure of equity. It is indisputable that there are huge inequities in the world: some of us enjoying staggering wealth while others have almost nothing at all. Billions of people remain without the basic necessities of a good life—clean water, safe housing, sanitation, and the wherewithal to earn a living for themselves and their families.[iv] Some of us have access to almost unimaginably advanced medical treatment; others have not even the rudiments of what we euphemistically call health care. Much of the world's wealth has come from exploiting resources in the global south to "satisfy markets in the industrialized north," and often this extraction has left a legacy of environmental damage. So there is an "ecological debt" that the wealthy owe to the poor, a debt that our current market-based policies and practices do not address.

To those who argue that the poor can best be served by expanding the reach of capitalism—that what we need is more unalloyed capitalism, not less—the pope insists that certain needs are not commodities to be bought and sold, but basic human rights.

> Even as the quality of available water is constantly diminishing, in some places there is a growing tendency . . . to privatize this resource, turning it into a commodity subject to the laws of the market. Yet *access to safe drinkable water is a basic and universal human right, since it is essential to human survival and, as such, is a condition for the exercise of other human rights.* Our world has a grave social debt towards the poor who lack access to drinking water, because *they are denied the right to a life consistent with their inalienable dignity.*

If the poor lack access to safe drinking water, the solution is not to charge them for it.

The second failure is environmental damage. The champions of our current system often say its benefits have simply not yet reached the poor—and therefore we must continue (and even strengthen) the practices that have made the rich rich until they reach all. This argument might be persuasive were it not for the damage that we have wreaked on the environment.

> It should always be kept in mind that "environmental protection cannot be assured solely on the basis of financial calculations of costs and benefits. The environment is one of those goods that cannot be adequately safeguarded or promoted by market forces." Once more, we need to reject a magical conception of the market, which would suggest that problems can be solved simply by an increase in the profits of companies or individuals. Is it realistic to hope that those who are obsessed with maximizing profits will stop to reflect on the environmental damage which they will leave behind for future generations? Where profits alone count, there can be no thinking about the rhythms of nature, its phases of decay and regeneration, or the complexity of ecosystems which may be gravely upset by human intervention. Moreover, biodiversity is considered at most a deposit of economic resources available for exploitation, with no serious thought for the real value of things, their significance for persons and cultures, or the concerns and needs of the poor.

Carbon markets have been widely advocated as the solution to climate change, but the pope has grave concerns here as well.

> The strategy of buying and selling "carbon credits" can lead to a new form of speculation which would not help reduce the emission of polluting gases worldwide. This system seems to provide . . . the guise of a certain commitment to the environment, but in no way does it allow for the radical change which present circumstances require. Rather, it may simply become a ploy which permits maintaining the excessive consumption of some countries and sectors.

Climate change, he suggests, is more than just the failure of markets to internalize the true cost of carbon. It is the failure of a system in which profit is "the sole criterion to be taken into account." After all, if we make fossil fuels more expensive to discourage their use, what happens to the poor, who need energy as much or more than the rich? Without a mechanism to help them gain access to clean energy, putting a price on carbon will only make their plight worse. Mechanisms to help the poor of course exist, but they involve reaching outside of markets and thinking beyond the profit motive.

Moreover, a great deal of environmental wreckage has been inflicted by multinational corporations operating in developing nations in ways that would not be acceptable in the developed world. When these companies cease their operations, frequently they leave substantial damage in their wake, damage that is not an accident or oversight but the consequence of an ideology in which "'whatever is fragile, like the environment, is defenceless before the interests of a deified market, which become the only rule.'"

The third failure is the spiritual impoverishment of the rich. The cheerleaders of capitalism insist that free markets are not just the best means of delivering goods and services, but the only means that protect our freedom.[v] In the aftermath of the Cold War, this can be a hard argument to refute, but the pope is a brave man and he takes on the challenge: Our paradigm leads people to believe that they are free "as long as they have the supposed freedom to consume." But the truly free are "the minority who wield economic and financial power." And are we really happy? He suspects not:

> Since the market tends to promote extreme consumerism in an effort to sell its products, people can easily get caught up in a whirlwind of needless buying and spending. Compulsive consumerism is one example of how the techno-economic paradigm affects individuals . . . Many people know that . . . the mere amassing of things and pleasures [is] not enough to give meaning and joy to the human heart, yet they feel unable to give up what the market sets before them.

The unhappiness of people in the wealthy west is perhaps the biggest indictment of our current system: even those of us who purportedly benefit from it do not seem very happy. Consider the bookstore shelves crammed with titles on happiness. A quick search on Amazon.com yields 409,429 results for "happiness," with subtitles like "How to Develop Life's Most Important Skill."[vi] How did we come to a place where we have such material riches, yet we need experts to tell us how to be happy? The pope knows: "We have too many means and only a few insubstantial ends."

The magic of the marketplace has turned out not to be so magical at all. Billions have not shared the wealth that markets have generated, those who have are not content, and meanwhile we have undermined the chances for happiness of future generations and even the chance for existence of many species. The "realism" of the status quo is one that denies the persistence and profundity of these failures. Thus the pope's two themes are in fact one: our failure to care for creation is the result of a world-view that defines everything in consumerist terms and looks for solutions solely in things that can be bought and sold. The market economy and the culture of consumption are locked in a dance of death, leading to spiritual impoverishment for those who control it, material impoverishment for those who don't, and environmental impoverishment across the globe.

THE ALTERNATIVE: AN INTEGRAL ECOLOGY

Some may shrug and say that environmental damage is the price of progress, but the pope refuses to accept that conclusion as *rational*. On the contrary, viewed dispassionately, it comes to look a bit insane. "The markets, which immediately benefit from sales, stimulate ever greater demand. An outsider looking at our world would be amazed at such behaviour, which at times appears self-destructive." The pope does not go so far as to label the technological paradigm religion, but his use of the phrase "deified market" certainly suggests that thought.

What is the alternative? This, of course, is the challenge—it is easy to see what is wrong with our system, harder to see how to fix it. The pope calls for an "integral ecology," by which he means a vision of the world founded fundamentally on respect for Creation, and a renewed emphasis on our mutual interconnection with one another and with nature in all its complexity. This ecology must necessarily include science and technology, but a science that moves past reductionism and a technology that is more focused on authentic needs. As the pope puts it: "We are faced not with two separate crises, one environmental and the other social, but rather with one complex crisis which is both social and environmental. Strategies for a solution demand an integrated approach to combating poverty, restoring dignity to the excluded, and at the same time protecting nature." It is not a question of development versus the environment, as was recognized at the Rio Earth Summit more than two decades ago, for the protection of the environment is "'an integral part of the development process and cannot be considered in isolation from it.'"

An integral ecology must also include also good governance, including "the establishment of a legal framework which can set clear boundaries and ensure the protection of ecosystems." It must replace a culture of rampant individualism with a culture of care rooted in "love for society and commitment to the common good." And it must not only reclaim the idea of the common good, but also recognize it as the centerpiece of civil society, environmental protection, religious communion, and, finally, human dignity, happiness, and love.

Crucially, this culture of care must include not just those of us alive today, but also future generations—a point the pope makes more than once, in both economic and moral terms. Our current economic models literally dis-count the future, insofar as damage in the future is counted as costing less than damage today, but what sort of a calculus is it that concludes that our needs are greater than our children's? The notion of the common good, the pope concludes, also extends to future generations: "We can no longer speak of sustainable development apart from intergenerational solidarity . . . Intergenerational solidarity is not optional, but rather a basic question of justice, since the world we have received also belongs to those who will follow us."

Some readers will be dissatisfied with this ending, in part because the logic of the technocratic paradigm is to insist there is no functional alternative. So perhaps this is the pope's most important message: that we must move past the ideology of no ideology, the morality of amorality.

> The idea of promoting a different cultural paradigm . . . is nowadays inconceivable. The technological paradigm has become so dominant that it would be difficult to do without its resources and even more difficult to utilize them without being dominated by their internal logic. It has become countercultural to choose a lifestyle whose goals are even partly independent of technology, of its costs and its power to globalize and make us all the same . . . Our capacity to make decisions, a more genuine freedom and the space for each one's alternative creativity are diminished.

Many people will of course misread this message. Already, conservatives are condemning the *Encyclical* for its failure to celebrate the rewards of capitalism and extol the virtues of carbon markets.[vii] Others have misread the letter depriving "people of the [technological] tools humanity will need to prevent climatic upheaval."[viii] These reactions demonstrate why this *Encyclical* is so important.

The pope is not asking us to reject markets or technology. He is asking us to reject the (il)logic that insists that only markets can decide our future and that technology is politically and morally neutral. He is asking us to reject the creed of market fundamentalism, and to recognize that the system has levers. Individuals, institutions, and governments are all making choices, and we have the capacity to make different ones.

Encyclical on Climate Change & Inequality

Preface

1. *"LAUDATO SI,' mi' Signore"*—*"Praise be to you, my Lord."* In the words of this beautiful canticle, Saint Francis of Assisi reminds us that our common home is like a sister with whom we share our life and a beautiful mother who opens her arms to embrace us. "Praise be to you, my Lord, through our Sister, Mother Earth, who sustains and governs us, and who produces various fruit with coloured flowers and herbs."[1]

2. This sister now cries out to us because of the harm we have inflicted on her by our irresponsible use and abuse of the goods with which God has endowed her. We have come to see ourselves as her lords and masters, entitled to plunder her at will. The violence present in our hearts, wounded by sin, is also reflected in the symptoms of sickness evident in the soil, in the water, in the air and in all forms of life. This is why the earth herself, burdened and laid waste, is among the most abandoned and maltreated of our poor; she "groans in travail"

(*Rom* 8:22). We have forgotten that we ourselves are dust of the
earth (cf. *Gen* 2:7); our very bodies are made up of her elements,
we breathe her air and we receive life and refreshment from
her waters.

NOTHING IN THIS WORLD IS INDIFFERENT TO US

3. More than fifty years ago, with the world teetering on the
brink of nuclear crisis, Pope Saint John XXIII wrote an Encyclical which not only rejected war but offered a proposal for
peace. He addressed his message *Pacem in Terris* to the entire
"Catholic world" and indeed "to all men and women of good
will." Now, faced as we are with global environmental deterioration, I wish to address every person living on this planet. In
my Apostolic Exhortation *Evangelii Gaudium*, I wrote to all the
members of the Church with the aim of encouraging ongoing
missionary renewal. In this Encyclical, I would like to enter
into dialogue with all people about our common home.

4. In 1971, eight years after *Pacem in Terris*, Blessed Pope Paul
VI referred to the ecological concern as "a tragic consequence"
of unchecked human activity: "Due to an ill-considered exploitation of nature, humanity runs the risk of destroying it and
becoming in turn a victim of this degradation."[2] He spoke in
similar terms to the Food and Agriculture Organization of the
United Nations about the potential for an "ecological catastro-
phe under the effective explosion of industrial civilization,"
and stressed "the urgent need for a radical change in the conduct of humanity," inasmuch as "the most extraordinary scien-

tific advances, the most amazing technical abilities, the most astonishing economic growth, unless they are accompanied by authentic social and moral progress, will definitively turn against man."[3]

5. Saint John Paul II became increasingly concerned about this issue. In his first Encyclical he warned that human beings frequently seem "to see no other meaning in their natural environment than what serves for immediate use and consumption."[4] Subsequently, he would call for a global ecological *conversion.*[5] At the same time, he noted that little effort had been made to "safeguard the moral conditions for an authentic *human ecology.*"[6] The destruction of the human environment is extremely serious, not only because God has entrusted the world to us men and women, but because human life is itself a gift which must be defended from various forms of debasement. Every effort to protect and improve our world entails profound changes in "lifestyles, models of production and consumption, and the established structures of power which today govern societies."[7] Authentic human development has a moral character. It presumes full respect for the human person, but it must also be concerned for the world around us and "take into account the nature of each being and of its mutual connection in an ordered system."[8] Accordingly, our human ability to transform reality must proceed in line with God's original gift of all that is.[9]

6. My predecessor Benedict XVI likewise proposed "eliminating the structural causes of the dysfunctions of the world economy and correcting models of growth which have proved incapable of ensuring respect for the environment."[10] He ob-

served that the world cannot be analyzed by isolating only one of its aspects, since "the book of nature is one and indivisible," and includes the environment, life, sexuality, the family, social relations, and so forth. It follows that "the deterioration of nature is closely connected to the culture which shapes human coexistence."[11] Pope Benedict asked us to recognize that the natural environment has been gravely damaged by our irresponsible behaviour. The social environment has also suffered damage. Both are ultimately due to the same evil: the notion that there are no indisputable truths to guide our lives, and hence human freedom is limitless. We have forgotten that "man is not only a freedom which he creates for himself. Man does not create himself. He is spirit and will, but also nature."[12] With paternal concern, Benedict urged us to realize that creation is harmed "where we ourselves have the final word, where everything is simply our property and we use it for ourselves alone. The misuse of creation begins when we no longer recognize any higher instance than ourselves, when we see nothing else but ourselves."[13]

UNITED BY THE SAME CONCERN

7. These statements of the Popes echo the reflections of numerous scientists, philosophers, theologians and civic groups, all of which have enriched the Church's thinking on these questions. Outside the Catholic Church, other Churches and Christian communities—and other religions as well—have expressed deep concern and offered valuable reflections on issues which all of us find disturbing. To give just one striking example, I

would mention the statements made by the beloved Ecumenical Patriarch Bartholomew, with whom we share the hope of full ecclesial communion.

8. Patriarch Bartholomew has spoken in particular of the need for each of us to repent of the ways we have harmed the planet, for "inasmuch as we all generate small ecological damage," we are called to acknowledge "our contribution, smaller or greater, to the disfigurement and destruction of creation."[14] He has repeatedly stated this firmly and persuasively, challenging us to acknowledge our sins against creation: "For human beings . . . to destroy the biological diversity of God's creation; for human beings to degrade the integrity of the earth by causing changes in its climate, by stripping the earth of its natural forests or destroying its wetlands; for human beings to contaminate the earth's waters, its land, its air, and its life—these are sins."[15] For "to commit a crime against the natural world is a sin against ourselves and a sin against God."[16]

9. At the same time, Bartholomew has drawn attention to the ethical and spiritual roots of environmental problems, which require that we look for solutions not only in technology but in a change of humanity; otherwise we would be dealing merely with symptoms. He asks us to replace consumption with sacrifice, greed with generosity, wastefulness with a spirit of sharing, an asceticism which "entails learning to give, and not simply to give up. It is a way of loving, of moving gradually away from what I want to what God's world needs. It is liberation from fear, greed and compulsion."[17] As Christians, we are also called "to accept the world as a sacrament of communion, as a way of

sharing with God and our neighbours on a global scale. It is our humble conviction that the divine and the human meet in the slightest detail in the seamless garment of God's creation, in the last speck of dust of our planet."[18]

SAINT FRANCIS OF ASSISI

10. I do not want to write this Encyclical without turning to that attractive and compelling figure, whose name I took as my guide and inspiration when I was elected Bishop of Rome. I believe that Saint Francis is the example par excellence of care for the vulnerable and of an integral ecology lived out joyfully and authentically. He is the patron saint of all who study and work in the area of ecology, and he is also much loved by non-Christians. He was particularly concerned for God's creation and for the poor and outcast. He loved, and was deeply loved for his joy, his generous self-giving, his openheartedness. He was a mystic and a pilgrim who lived in simplicity and in wonderful harmony with God, with others, with nature and with himself. He shows us just how inseparable the bond is between concern for nature, justice for the poor, commitment to society and interior peace.

11. Francis helps us to see that an integral ecology calls for openness to categories which transcend the language of mathematics and biology, and take us to the heart of what it is to be human. Just as happens when we fall in love with someone, whenever he would gaze at the sun, the moon or the smallest of animals, he burst into song, drawing all other creatures into

his praise. He communed with all creation, even preaching to the flowers, inviting them "to praise the Lord, just as if they were endowed with reason."[19] His response to the world around him was so much more than intellectual appreciation or economic calculus, for to him each and every creature was a sister united to him by bonds of affection. That is why he felt called to care for all that exists. His disciple Saint Bonaventure tells us that, "from a reflection on the primary source of all things, filled with even more abundant piety, he would call creatures, no matter how small, by the name of 'brother' or 'sister.'"[20] Such a conviction cannot be written off as naive romanticism, for it affects the choices which determine our behaviour. If we approach nature and the environment without this openness to awe and wonder, if we no longer speak the language of fraternity and beauty in our relationship with the world, our attitude will be that of masters, consumers, ruthless exploiters, unable to set limits on their immediate needs. By contrast, if we feel intimately united with all that exists, then sobriety and care will well up spontaneously. The poverty and austerity of Saint Francis were no mere veneer of asceticism, but something much more radical: a refusal to turn reality into an object simply to be used and controlled.

12. What is more, Saint Francis, faithful to Scripture, invites us to see nature as a magnificent book in which God speaks to us and grants us a glimpse of his infinite beauty and goodness. "Through the greatness and the beauty of creatures one comes to know by analogy their maker" (*Wis* 13:5); indeed, "his eternal power and divinity have been made known through his works since the creation of the world" (*Rom* 1:20). For this reason,

Francis asked that part of the friary garden always be left untouched, so that wild flowers and herbs could grow there, and those who saw them could raise their minds to God, the Creator of such beauty.[21] Rather than a problem to be solved, the world is a joyful mystery to be contemplated with gladness and praise.

MY APPEAL

13. The urgent challenge to protect our common home includes a concern to bring the whole human family together to seek a sustainable and integral development, for we know that things can change. The Creator does not abandon us; he never forsakes his loving plan or repents of having created us. Humanity still has the ability to work together in building our common home. Here I want to recognize, encourage and thank all those striving in countless ways to guarantee the protection of the home which we share. Particular appreciation is owed to those who tirelessly seek to resolve the tragic effects of environmental degradation on the lives of the world's poorest. Young people demand change. They wonder how anyone can claim to be building a better future without thinking of the environmental crisis and the sufferings of the excluded.

14. I urgently appeal, then, for a new dialogue about how we are shaping the future of our planet. We need a conversation which includes everyone, since the environmental challenge we are undergoing, and its human roots, concern and affect us all. The worldwide ecological movement has already made considerable

progress and led to the establishment of numerous organizations committed to raising awareness of these challenges.
Regrettably, many efforts to seek concrete solutions to the environmental crisis have proved ineffective, not only because of
powerful opposition but also because of a more general lack of
interest. Obstructionist attitudes, even on the part of believers,
can range from denial of the problem to indifference, noncha-
lant resignation or blind confidence in technical solutions. We
require a new and universal solidarity. As the bishops of Southern Africa have stated: "Everyone's talents and involvement are
needed to redress the damage caused by human abuse of God's
creation."[22] All of us can cooperate as instruments of God for
the care of creation, each according to his or her own culture,
experience, involvements and talents.

15. It is my hope that this Encyclical Letter, which is now added
to the body of the Church's social teaching, can help us to acknowledge the appeal, immensity and urgency of the challenge
we face. I will begin by briefly reviewing several aspects of the
present ecological crisis, with the aim of drawing on the results
of the best scientific research available today, letting them touch
us deeply and provide a concrete foundation for the ethical and
spiritual itinerary that follows. I will then consider some principles drawn from the Judaeo-Christian tradition which can
render our commitment to the environment more coherent. I
will then attempt to get to the roots of the present situation, so
as to consider not only its symptoms but also its deepest causes.
This will help to provide an approach to ecology which respects
our unique place as human beings in this world and our relationship to our surroundings. In light of this reflection, I will

advance some broader proposals for dialogue and action which would involve each of us as individuals, and also affect international policy. Finally, convinced as I am that change is impossible without motivation and a process of education, I will offer some inspired guidelines for human development to be found in the treasure of Christian spiritual experience.

16. Although each chapter will have its own subject and specific approach, it will also take up and re-examine important questions previously dealt with. This is particularly the case with a number of themes which will reappear as the Encyclical unfolds. As examples, I will point to the intimate relationship between the poor and the fragility of the planet, the conviction that everything in the world is connected, the critique of new paradigms and forms of power derived from technology, the call to seek other ways of understanding the economy and progress, the value proper to each creature, the human meaning of ecology, the need for forthright and honest debate, the serious responsibility of international and local policy, the throwaway culture and the proposal of a new lifestyle. These questions will not be dealt with once and for all, but reframed and enriched again and again.

CHAPTER ONE

What Is Happening to Our Common Home

17. Theological and philosophical reflections on the situation of humanity and the world can sound tiresome and abstract, unless they are grounded in a fresh analysis of our present situation, which is in many ways unprecedented in the history of humanity. So, before considering how faith brings new incentives and requirements with regard to the world of which we are a part, I will briefly turn to what is happening to our common home.

18. The continued acceleration of changes affecting humanity and the planet is coupled today with a more intensified pace of life and work which might be called "rapidification." Although change is part of the working of complex systems, the speed with which human activity has developed contrasts with the naturally slow pace of biological evolution. Moreover, the goals of this rapid and constant change are not necessarily geared to

the common good or to integral and sustainable human development. Change is something desirable, yet it becomes a source of anxiety when it causes harm to the world and to the quality of life of much of humanity.

19. Following a period of irrational confidence in progress and human abilities, some sectors of society are now adopting a more critical approach. We see increasing sensitivity to the environment and the need to protect nature, along with a growing concern, both genuine and distressing, for what is happening to our planet. Let us review, however cursorily, those questions which are troubling us today and which we can no longer sweep under the carpet. Our goal is not to amass information or to satisfy curiosity, but rather to become painfully aware, to dare to turn what is happening to the world into our own personal suffering and thus to discover what each of us can do about it.

I. Pollution and Climate Change

POLLUTION, WASTE AND THE THROWAWAY CULTURE

20. Some forms of pollution are part of people's daily experience. Exposure to atmospheric pollutants produces a broad spectrum of health hazards, especially for the poor, and causes millions of premature deaths. People take sick, for example, from breathing high levels of smoke from fuels used in cooking or heating. There is also pollution that affects everyone, caused

by transport, industrial fumes, substances which contribute to the acidification of soil and water, fertilizers, insecticides, fungicides, herbicides and agrotoxins in general. Technology, which, linked to business interests, is presented as the only way of solving these problems, in fact proves incapable of seeing the mysterious network of relations between things and so sometimes solves one problem only to create others.

21. Account must also be taken of the pollution produced by residue, including dangerous waste present in different areas. Each year hundreds of millions of tons of waste are generated, much of it non-biodegradable, highly toxic and radioactive, from homes and businesses, from construction and demolition sites, from clinical, electronic and industrial sources. The earth, our home, is beginning to look more and more like an immense pile of filth. In many parts of the planet, the elderly lament that once beautiful landscapes are now covered with rubbish. Industrial waste and chemical products utilized in cities and agricultural areas can lead to bioaccumulation in the organisms of the local population, even when levels of toxins in those places are low. Frequently no measures are taken until after people's health has been irreversibly affected.

22. These problems are closely linked to a throwaway culture which affects the excluded just as it quickly reduces things to rubbish. To cite one example, most of the paper we produce is thrown away and not recycled. It is hard for us to accept that the way natural ecosystems work is exemplary: plants synthesize nutrients which feed herbivores; these in turn become food for carnivores, which produce significant quantities of organic

waste which give rise to new generations of plants. But our industrial system, at the end of its cycle of production and consumption, has not developed the capacity to absorb and reuse
waste and by-products. We have not yet managed to adopt a
circular model of production capable of preserving resources
for present and future generations, while limiting as much as
possible the use of non-renewable resources, moderating their
consumption, maximizing their efficient use, reusing and recycling them. A serious consideration of this issue would be one
way of counteracting the throwaway culture which affects the
entire planet, but it must be said that only limited progress has
been made in this regard.

CLIMATE AS A COMMON GOOD

23. The climate is a common good, belonging to all and meant
for all. At the global level, it is a complex system linked to many
of the essential conditions for human life. A very solid scientific
consensus indicates that we are presently witnessing a disturbing
warming of the climatic system. In recent decades this warming has been accompanied by a constant rise in the sea level
and, it would appear, by an increase of extreme weather events,
even if a scientifically determinable cause cannot be assigned to
each particular phenomenon. Humanity is called to recognize
the need for changes of lifestyle, production and consumption,
in order to combat this warming or at least the human causes
which produce or aggravate it. It is true that there are other
factors (such as volcanic activity, variations in the earth's or-
bit and axis, the solar cycle), yet a number of scientific studies

indicate that most global warming in recent decades is due to the great concentration of greenhouse gases (carbon dioxide, methane, nitrogen oxides and others) released mainly as a result of human activity. Concentrated in the atmosphere, these gases do not allow the warmth of the sun's rays reflected by the earth to be dispersed in space. The problem is aggravated by a model of development based on the intensive use of fossil fuels, which is at the heart of the worldwide energy system. Another determining factor has been an increase in changed uses of the soil, principally deforestation for agricultural purposes.

24. Warming has effects on the carbon cycle. It creates a vicious circle which aggravates the situation even more, affecting the availability of essential resources like drinking water, energy and agricultural production in warmer regions, and leading to the extinction of part of the planet's biodiversity. The melting in the polar ice caps and in high altitude plains can lead to the dangerous release of methane gas, while the decomposition of frozen organic material can further increase the emission of carbon dioxide. Things are made worse by the loss of tropical forests which would otherwise help to mitigate climate change. Carbon dioxide pollution increases the acidification of the oceans and compromises the marine food chain. If present trends continue, this century may well witness extraordinary climate change and an unprecedented destruction of ecosystems, with serious consequences for all of us. A rise in the sea level, for example, can create extremely serious situations, if we consider that a quarter of the world's population lives on the coast or nearby, and that the majority of our megacities are situated in coastal areas.

25. Climate change is a global problem with grave implications: environmental, social, economic, political and for the distribution of goods. It represents one of the principal challenges facing humanity in our day. Its worst impact will probably be felt by developing countries in coming decades. Many of the poor live in areas particularly affected by phenomena related to warming, and their means of subsistence are largely dependent on natural reserves and ecosystemic services such as agriculture, fishing and forestry. They have no other financial activities or resources which can enable them to adapt to climate change or to face natural disasters, and their access to social services and protection is very limited. For example, changes in climate, to which animals and plants cannot adapt, lead them to migrate; this in turn affects the livelihood of the poor, who are then forced to leave their homes, with great uncertainty for their future and that of their children. There has been a tragic rise in the number of migrants seeking to flee from the growing poverty caused by environmental degradation. They are not recognized by international conventions as refugees; they bear the loss of the lives they have left behind, without enjoying any legal protection whatsoever. Sadly, there is widespread indifference to such suffering, which is even now taking place throughout our world. Our lack of response to these tragedies involving our brothers and sisters points to the loss of that sense of responsibility for our fellow men and women upon which all civil society is founded.

26. Many of those who possess more resources and economic or political power seem mostly to be concerned with masking the problems or concealing their symptoms, simply making efforts

to reduce some of the negative impacts of climate change. However, many of these symptoms indicate that such effects will continue to worsen if we continue with current models of production and consumption. There is an urgent need to develop policies so that, in the next few years, the emission of carbon dioxide and other highly polluting gases can be drastically reduced, for example, substituting for fossil fuels and developing sources of renewable energy. Worldwide there is minimal access to clean and renewable energy. There is still a need to develop adequate storage technologies. Some countries have made considerable progress, although it is far from constituting a significant proportion. Investments have also been made in means of production and transportation which consume less energy and require fewer raw materials, as well as in methods of construction and renovating buildings which improve their energy efficiency. But these good practices are still far from widespread.

II. The Issue of Water

27. Other indicators of the present situation have to do with the depletion of natural resources. We all know that it is not possible to sustain the present level of consumption in developed countries and wealthier sectors of society, where the habit of wasting and discarding has reached unprecedented levels. The exploitation of the planet has already exceeded acceptable limits and we still have not solved the problem of poverty.

28. Fresh drinking water is an issue of primary importance, since it is indispensable for human life and for supporting terrestrial and aquatic ecosystems. Sources of fresh water are necessary for health care, agriculture and industry. Water supplies used to be relatively constant, but now in many places demand exceeds the sustainable supply, with dramatic consequences in the short and long term. Large cities dependent on significant supplies of water have experienced periods of shortage, and at critical moments these have not always been administered with sufficient oversight and impartiality. Water poverty especially affects Africa where large sectors of the population have no access to safe drinking water or experience droughts which impede agricultural production. Some countries have areas rich in water while others endure drastic scarcity.

29. One particularly serious problem is the quality of water available to the poor. Every day, unsafe water results in many deaths and the spread of water-related diseases, including those caused by microorganisms and chemical substances. Dysentery and cholera, linked to inadequate hygiene and water supplies, are a significant cause of suffering and of infant mortality. Underground water sources in many places are threatened by the pollution produced in certain mining, farming and industrial activities, especially in countries lacking adequate regulation or controls. It is not only a question of industrial waste. Detergents and chemical products, commonly used in many places of the world, continue to pour into our rivers, lakes and seas.

30. Even as the quality of available water is constantly diminishing, in some places there is a growing tendency, despite its

scarcity, to privatize this resource, turning it into a commodity subject to the laws of the market. Yet *access to safe drinkable water is a basic and universal human right, since it is essential to human survival and, as such, is a condition for the exercise of other human rights.* Our world has a grave social debt towards the poor who lack access to drinking water, because *they are denied the right to a life consistent with their inalienable dignity.* This debt can be paid partly by an increase in funding to provide clean water and sanitary services among the poor. But water continues to be wasted, not only in the developed world but also in developing countries which possess it in abundance. This shows that the problem of water is partly an educational and cultural issue, since there is little awareness of the seriousness of such behaviour within a context of great inequality.

31. Greater scarcity of water will lead to an increase in the cost of food and the various products which depend on its use. Some studies warn that an acute water shortage may occur within a few decades unless urgent action is taken. The environmental repercussions could affect billions of people; it is also conceivable that the control of water by large multinational businesses may become a major source of conflict in this century.[23]

III. Loss of Biodiversity

32. The earth's resources are also being plundered because of short-sighted approaches to the economy, commerce and production. The loss of forests and woodlands entails the loss of

species which may constitute extremely important resources in the future, not only for food but also for curing disease and other uses. Different species contain genes which could be key resources in years ahead for meeting human needs and regulating environmental problems.

33. It is not enough, however, to think of different species merely as potential "resources" to be exploited, while overlooking the fact that they have value in themselves. Each year sees the disappearance of thousands of plant and animal species which we will never know, which our children will never see, because they have been lost for ever. The great majority become extinct for reasons related to human activity. Because of us, thousands of species will no longer give glory to God by their very existence, nor convey their message to us. We have no such right.

34. It may well disturb us to learn of the extinction of mammals or birds, since they are more visible. But the good functioning of ecosystems also requires fungi, algae, worms, insects, reptiles and an innumerable variety of microorganisms. Some less numerous species, although generally unseen, nonetheless play a critical role in maintaining the equilibrium of a particular place. Human beings must intervene when a geosystem reaches a critical state. But nowadays, such intervention in nature has become more and more frequent. As a consequence, serious problems arise, leading to further interventions; human activity becomes ubiquitous, with all the risks which this entails. Often a vicious circle results, as human intervention to resolve a problem further aggravates the situation. For example, many

birds and insects which disappear due to synthetic agrotoxins are helpful for agriculture: their disappearance will have to be compensated for by yet other techniques which may well prove harmful. We must be grateful for the praiseworthy efforts being made by scientists and engineers dedicated to finding solutions to man-made problems. But a sober look at our world shows that the degree of human intervention, often in the service of business interests and consumerism, is actually making our earth less rich and beautiful, ever more limited and grey, even as technological advances and consumer goods continue to abound limitlessly. We seem to think that we can substitute an irreplaceable and irretrievable beauty with something which we have created ourselves.

35. In assessing the environmental impact of any project, concern is usually shown for its effects on soil, water and air, yet few careful studies are made of its impact on biodiversity, as if the loss of species or animals and plant groups were of little importance. Highways, new plantations, the fencing-off of certain areas, the damming of water sources, and similar developments, crowd out natural habitats and, at times, break them up in such a way that animal populations can no longer migrate or roam freely. As a result, some species face extinction. Alternatives exist which at least lessen the impact of these projects, like the creation of biological corridors, but few countries demonstrate such concern and foresight. Frequently, when certain species are exploited commercially, little attention is paid to studying their reproductive patterns in order to prevent their depletion and the consequent imbalance of the ecosystem.

36. Caring for ecosystems demands far-sightedness, since no one looking for quick and easy profit is truly interested in their preservation. But the cost of the damage caused by such selfish lack of concern is much greater than the economic benefits to be obtained. Where certain species are destroyed or seriously harmed, the values involved are incalculable. We can be silent witnesses to terrible injustices if we think that we can obtain significant benefits by making the rest of humanity, present and future, pay the extremely high costs of environmental deterioration.

37. Some countries have made significant progress in establishing sanctuaries on land and in the oceans where any human intervention is prohibited which might modify their features or alter their original structures. In the protection of biodiversity, specialists insist on the need for particular attention to be shown to areas richer both in the number of species and in endemic, rare or less protected species. Certain places need greater protection because of their immense importance for the global ecosystem, or because they represent important water reserves and thus safeguard other forms of life.

38. Let us mention, for example, those richly biodiverse lungs of our planet which are the Amazon and the Congo basins, or the great aquifers and glaciers. We know how important these are for the entire earth and for the future of humanity. The ecosystems of tropical forests possess an enormously complex biodiversity which is almost impossible to appreciate fully, yet when these forests are burned down or levelled for purposes of cultivation, within the space of a few years countless species are

lost and the areas frequently become arid wastelands. A delicate balance has to be maintained when speaking about these
places, for we cannot overlook the huge global economic interests which, under the guise of protecting them, can under-
mine the sovereignty of individual nations. In fact, there are
"proposals to internationalize the Amazon, which only serve
the economic interests of transnational corporations."[24] We
cannot fail to praise the commitment of international agencies
and civil society organizations which draw public attention to
these issues and offer critical cooperation, employing legitimate
means of pressure, to ensure that each government carries out
its proper and inalienable responsibility to preserve its country's environment and natural resources, without capitulating
to spurious local or international interests.

39. The replacement of virgin forest with plantations of trees,
usually monocultures, is rarely adequately analyzed. Yet this
can seriously compromise a biodiversity which the new species
being introduced does not accommodate. Similarly, wetlands
converted into cultivated land lose the enormous biodiversity
which they formerly hosted. In some coastal areas the disappearance of ecosystems sustained by mangrove swamps is a
source of serious concern.

40. Oceans not only contain the bulk of our planet's water supply, but also most of the immense variety of living creatures,
many of them still unknown to us and threatened for various
reasons. What is more, marine life in rivers, lakes, seas and
oceans, which feeds a great part of the world's population, is affected by uncontrolled fishing, leading to a drastic depletion of

certain species. Selective forms of fishing which discard much of what they collect continue unabated. Particularly threatened are marine organisms which we tend to overlook, like some forms of plankton; they represent a significant element in the ocean food chain, and species used for our food ultimately depend on them.

41. In tropical and subtropical seas, we find coral reefs comparable to the great forests on dry land, for they shelter approximately a million species, including fish, crabs, molluscs, sponges and algae. Many of the world's coral reefs are already barren or in a state of constant decline. "Who turned the wonderworld of the seas into underwater cemeteries bereft of colour and life?"[25] This phenomenon is due largely to pollution which reaches the sea as the result of deforestation, agricultural monocultures, industrial waste and destructive fishing methods, especially those using cyanide and dynamite. It is aggravated by the rise in temperature of the oceans. All of this helps us to see that every intervention in nature can have consequences which are not immediately evident, and that certain ways of exploiting resources prove costly in terms of degradation which ultimately reaches the ocean bed itself.

42. Greater investment needs to be made in research aimed at understanding more fully the functioning of ecosystems and adequately analyzing the different variables associated with any significant modification of the environment. Because all creatures are connected, each must be cherished with love and respect, for all of us as living creatures are dependent on one another. Each area is responsible for the care of this family.

This will require undertaking a careful inventory of the species which it hosts, with a view to developing programmes and strategies of protection with particular care for safeguarding species heading towards extinction.

IV. Decline in the Quality of Human Life and the Breakdown of Society

43. Human beings too are creatures of this world, enjoying a right to life and happiness, and endowed with unique dignity. So we cannot fail to consider the effects on people's lives of environmental deterioration, current models of development and the throwaway culture.

44. Nowadays, for example, we are conscious of the disproportionate and unruly growth of many cities, which have become unhealthy to live in, not only because of pollution caused by toxic emissions but also as a result of urban chaos, poor transportation, and visual pollution and noise. Many cities are huge, inefficient structures, excessively wasteful of energy and water. Neighbourhoods, even those recently built, are congested, chaotic and lacking in sufficient green space. We were not meant to be inundated by cement, asphalt, glass and metal, and deprived of physical contact with nature.

45. In some places, rural and urban alike, the privatization of certain spaces has restricted people's access to places of partic-

ular beauty. In others, "ecological" neighbourhoods have been created which are closed to outsiders in order to ensure an artificial tranquillity. Frequently, we find beautiful and carefully manicured green spaces in so-called "safer" areas of cities, but not in the more hidden areas where the disposable of society live.

46. The social dimensions of global change include the effects of technological innovations on employment, social exclusion, an inequitable distribution and consumption of energy and other services, social breakdown, increased violence and a rise in new forms of social aggression, drug trafficking, growing drug use by young people, and the loss of identity. These are signs that the growth of the past two centuries has not always led to an integral development and an improvement in the quality of life. Some of these signs are also symptomatic of real social decline, the silent rupture of the bonds of integration and social cohesion.

47. Furthermore, when media and the digital world become omnipresent, their influence can stop people from learning how to live wisely, to think deeply and to love generously. In this context, the great sages of the past run the risk of going unheard amid the noise and distractions of an information overload. Efforts need to be made to help these media become sources of new cultural progress for humanity and not a threat to our deepest riches. True wisdom, as the fruit of self-examination, dialogue and generous encounter between persons, is not acquired by a mere accumulation of data which eventually leads to overload and confusion, a sort of mental pollution. Real

relationships with others, with all the challenges they entail, now tend to be replaced by a type of internet communication which enables us to choose or eliminate relationships at whim, thus giving rise to a new type of contrived emotion which has more to do with devices and displays than with other people and with nature. Today's media do enable us to communicate and to share our knowledge and affections. Yet at times they also shield us from direct contact with the pain, the fears and the joys of others and the complexity of their personal experiences. For this reason, we should be concerned that, alongside the exciting possibilities offered by these media, a deep and melancholic dissatisfaction with interpersonal relations, or a harmful sense of isolation, can also arise.

V. Global Inequality

48. The human environment and the natural environment deteriorate together; we cannot adequately combat environmental degradation unless we attend to causes related to human and social degradation. In fact, the deterioration of the environment and of society affects the most vulnerable people on the planet: "Both everyday experience and scientific research show that the gravest effects of all attacks on the environment are suffered by the poorest."[26] For example, the depletion of fishing reserves especially hurts small fishing communities without the means to replace those resources; water pollution particularly affects the poor who cannot buy bottled water; and rises in the sea level mainly affect impoverished coastal populations who

have nowhere else to go. The impact of present imbalances is also seen in the premature death of many of the poor, in conflicts sparked by the shortage of resources, and in any number of other problems which are insufficiently represented on global agendas.[27]

49. It needs to be said that, generally speaking, there is little in the way of clear awareness of problems which especially affect the excluded. Yet they are the majority of the planet's population, billions of people. These days, they are mentioned in international political and economic discussions, but one often has the impression that their problems are brought up as an afterthought, a question which gets added almost out of duty or in a tangential way, if not treated merely as collateral damage. Indeed, when all is said and done, they frequently remain at the bottom of the pile. This is due partly to the fact that many professionals, opinion makers, communications media and centres of power, being located in affluent urban areas, are far removed from the poor, with little direct contact with their problems. They live and reason from the comfortable position of a high level of development and a quality of life well beyond the reach of the majority of the world's population. This lack of physical contact and encounter, encouraged at times by the disintegration of our cities, can lead to a numbing of conscience and to tendentious analyses which neglect parts of reality. At times this attitude exists side by side with a "green" rhetoric. Today, however, we have to realize that a true ecological approach *always* becomes a social approach; it must integrate questions of justice in debates on the environment, so as to hear *both the cry of the earth and the cry of the poor.*

50. Instead of resolving the problems of the poor and thinking of how the world can be different, some can only propose a reduction in the birth rate. At times, developing countries face forms of international pressure which make economic assistance contingent on certain policies of "reproductive health." Yet "while it is true that an unequal distribution of the population and of available resources creates obstacles to development and a sustainable use of the environment, it must nonetheless be recognized that demographic growth is fully compatible with an integral and shared development."[28] To blame population growth instead of extreme and selective consumerism on the part of some, is one way of refusing to face the issues. It is an attempt to legitimize the present model of distribution, where a minority believes that it has the right to consume in a way which can never be universalized, since the planet could not even contain the waste products of such consumption. Besides, we know that approximately a third of all food produced is discarded, and "whenever food is thrown out it is as if it were stolen from the table of the poor."[29] Still, attention needs to be paid to imbalances in population density, on both national and global levels, since a rise in consumption would lead to complex regional situations, as a result of the interplay between problems linked to environmental pollution, transport, waste treatment, loss of resources and quality of life.

51. Inequity affects not only individuals but entire countries; it compels us to consider an ethics of international relations. A true "ecological debt" exists, particularly between the global north and south, connected to commercial imbalances with effects on the environment, and the disproportionate use of

natural resources by certain countries over long periods of time. The export of raw materials to satisfy markets in the industrialized north has caused harm locally, as for example in mercury pollution in gold mining or sulphur dioxide pollution in copper mining. There is a pressing need to calculate the use of environmental space throughout the world for depositing gas residues which have been accumulating for two centuries and have created a situation which currently affects all the countries of the world. The warming caused by huge consumption on the part of some rich countries has repercussions on the poorest areas of the world, especially Africa, where a rise in temperature, together with drought, has proved devastating for farming. There is also the damage caused by the export of solid waste and toxic liquids to developing countries, and by the pollution produced by companies which operate in less developed countries in ways they could never do at home, in the countries in which they raise their capital: "We note that often the .businesses which operate this way are multinationals. They do here what they would never do in developed countries or the so-called first world. Generally, after ceasing their activity and withdrawing, they leave behind great human and environmental liabilities such as unemployment, abandoned towns, the depletion of natural reserves, deforestation, the impoverishment of agriculture and local stock breeding, open pits, riven hills, polluted rivers and a handful of social works which are no longer sustainable."[30]

52. The foreign debt of poor countries has become a way of controlling them, yet this is not the case where ecological debt is concerned. In different ways, developing countries, where the

most important reserves of the biosphere are found, continue to fuel the development of richer countries at the cost of their own present and future. The land of the southern poor is rich and mostly unpolluted, yet access to ownership of goods and resources for meeting vital needs is inhibited by a system of commercial relations and ownership which is structurally perverse. The developed countries ought to help pay this debt by significantly limiting their consumption of non-renewable energy and by assisting poorer countries to support policies and programmes of sustainable development. The poorest areas and countries are less capable of adopting new models for reducing environmental impact because they lack the wherewithal to develop the necessary processes and to cover their costs. We must continue to be aware that, regarding climate change, there are *differentiated responsibilities*. As the United States bishops have said, greater attention must be given to "the needs of the poor, the weak and the vulnerable, in a debate often dominated by more powerful interests."[31] We need to strengthen the conviction that we are one single human family. There are no frontiers or barriers, political or social, behind which we can hide, still less is there room for the globalization of indifference.

VI. Weak Responses

53. These situations have caused sister earth, along with all the abandoned of our world, to cry out, pleading that we take another course. Never have we so hurt and mistreated our

common home as we have in the last two hundred years. Yet we are called to be instruments of God our Father, so that our planet might be what he desired when he created it and correspond with his plan for peace, beauty and fullness. The problem is that we still lack the culture needed to confront this crisis. We lack leadership capable of striking out on new paths and meeting the needs of the present with concern for all and without prejudice towards coming generations. The establishment of a legal framework which can set clear boundaries and ensure the protection of ecosystems has become indispensable; otherwise, the new power structures based on the techno-economic paradigm may overwhelm not only our politics but also freedom and justice.

54. It is remarkable how weak international political responses have been. The failure of global summits on the environment makes it plain that our politics are subject to technology and finance. There are too many special interests, and economic interests easily end up trumping the common good and manipulating information so that their own plans will not be affected. The *Aparecida Document* urges that "the interests of economic groups which irrationally demolish sources of life should not prevail in dealing with natural resources."[32] The alliance between the economy and technology ends up sidelining anything unrelated to its immediate interests. Consequently the most one can expect is superficial rhetoric, sporadic acts of philanthropy and perfunctory expressions of concern for the environment, whereas any genuine attempt by groups within society to introduce change is viewed as a nuisance based on romantic illusions or an obstacle to be circumvented.

55. Some countries are gradually making significant progress, developing more effective controls and working to combat corruption. People may well have a growing ecological sensitivity but it has not succeeded in changing their harmful habits of consumption which, rather than decreasing, appear to be growing all the more. A simple example is the increasing use and power of air-conditioning. The markets, which immediately benefit from sales, stimulate ever greater demand. An outsider looking at our world would be amazed at such behaviour, which at times appears self-destructive.

56. In the meantime, economic powers continue to justify the current global system where priority tends to be given to speculation and the pursuit of financial gain, which fail to take the context into account, let alone the effects on human dignity and the natural environment. Here we see how environmental deterioration and human and ethical degradation are closely linked. Many people will deny doing anything wrong because distractions constantly dull our consciousness of just how limited and finite our world really is. As a result, "whatever is fragile, like the environment, is defenceless before the interests of a deified market, which become the only rule."[33]

57. It is foreseeable that, once certain resources have been depleted, the scene will be set for new wars, albeit under the guise of noble claims. War always does grave harm to the environment and to the cultural riches of peoples, risks which are magnified when one considers nuclear arms and biological weapons. "Despite the international agreements which prohibit chemical, bacteriological and biological warfare, the fact is that

laboratory research continues to develop new offensive weapons
capable of altering the balance of nature."³⁴ Politics must pay
greater attention to foreseeing new conflicts and addressing the
causes which can lead to them. But powerful financial interests
prove most resistant to this effort, and political planning tends
to lack breadth of vision. What would induce anyone, at this
stage, to hold on to power only to be remembered for their inability to take action when it was urgent and necessary to do so?

58. In some countries, there are positive examples of environmental improvement: rivers, polluted for decades, have been
cleaned up; native woodlands have been restored; landscapes
have been beautified thanks to environmental renewal projects; beautiful buildings have been erected; advances have been
made in the production of non-polluting energy and in the improvement of public transportation. These achievements do not
solve global problems, but they do show that men and women
are still capable of intervening positively. For all our limitations, gestures of generosity, solidarity and care cannot but well
up within us, since we were made for love.

59. At the same time we can note the rise of a false or superficial
ecology which bolsters complacency and a cheerful reckless-
ness. As often occurs in periods of deep crisis which require
bold decisions, we are tempted to think that what is happening is not entirely clear. Superficially, apart from a few obvious
signs of pollution and deterioration, things do not look that serious, and the planet could continue as it is for some time. Such
evasiveness serves as a licence to carrying on with our present
lifestyles and models of production and consumption. This is

the way human beings contrive to feed their self-destructive vices: trying not to see them, trying not to acknowledge them, delaying the important decisions and pretending that nothing will happen.

VII. A Variety of Opinions

60. Finally, we need to acknowledge that different approaches and lines of thought have emerged regarding this situation and its possible solutions. At one extreme, we find those who doggedly uphold the myth of progress and tell us that ecological problems will solve themselves simply with the application of new technology and without any need for ethical considerations or deep change. At the other extreme are those who view men and women and all their interventions as no more than a threat, jeopardizing the global ecosystem, and consequently the presence of human beings on the planet should be reduced and all forms of intervention prohibited. Viable future scenarios will have to be generated between these extremes, since there is no one path to a solution. This makes a variety of proposals possible, all capable of entering into dialogue with a view to developing comprehensive solutions.

61. On many concrete questions, the Church has no reason to offer a definitive opinion; she knows that honest debate must be encouraged among experts, while respecting divergent views. But we need only take a frank look at the facts to see that our common home is falling into serious disrepair. Hope would

have us recognize that there is always a way out, that we can always redirect our steps, that we can always do something to solve our problems. Still, we can see signs that things are now reaching a breaking point, due to the rapid pace of change and degradation; these are evident in large-scale natural disasters as well as social and even financial crises, for the world's problems cannot be analyzed or explained in isolation. There are regions now at high risk and, aside from all doomsday predictions, the present world system is certainly unsustainable from a number of points of view, for we have stopped thinking about the goals of human activity. "If we scan the regions of our planet, we immediately see that humanity has disappointed God's expectations."[35]

CHAPTER TWO

The Gospel of Creation

62. Why should this document, addressed to all people of good will, include a chapter dealing with the convictions of believers? I am well aware that in the areas of politics and philosophy there are those who firmly reject the idea of a Creator, or consider it irrelevant, and consequently dismiss as irrational the rich contribution which religions can make towards an integral ecology and the full development of humanity. Others view religions simply as a subculture to be tolerated. Nonetheless, science and religion, with their distinctive approaches to understanding reality, can enter into an intense dialogue fruitful for both.

I. The Light Offered by Faith

63. Given the complexity of the ecological crisis and its multiple causes, we need to realize that the solutions will not

emerge from just one way of interpreting and transforming reality. Respect must also be shown for the various cultural riches of different peoples, their art and poetry, their interior life and spirituality. If we are truly concerned to develop an ecology capable of remedying the damage we have done, no branch of the sciences and no form of wisdom can be left out, and that includes religion and the language particular to it. The Catholic Church is open to dialogue with philosophical thought; this has enabled her to produce various syntheses between faith and reason. The development of the Church's social teaching represents such a synthesis with regard to social issues; this teaching is called to be enriched by taking up new challenges.

64. Furthermore, although this Encyclical welcomes dialogue with everyone so that together we can seek paths of liberation, I would like from the outset to show how faith convictions can offer Christians, and some other believers as well, ample motivation to care for nature and for the most vulnerable of their brothers and sisters. If the simple fact of being human moves people to care for the environment of which they are a part, Christians in their turn "realize that their responsibility within creation, and their duty towards nature and the Creator, are an essential part of their faith."[36] It is good for humanity and the world at large when we believers better recognize the ecological commitments which stem from our convictions.

II. The Wisdom of
the Biblical Accounts

65. Without repeating the entire theology of creation, we can
ask what the great biblical narratives say about the relationship
of human beings with the world. In the first creation account
in the Book of Genesis, God's plan includes creating humanity. After the creation of man and woman, "God saw everything that he had made, and behold it was *very good*" (*Gen* 1:31).
The Bible teaches that every man and woman is created out of
love and made in God's image and likeness (cf. *Gen* 1:26). This
shows us the immense dignity of each person, "who is not just
something, but someone. He is capable of self-knowledge, of
self-possession and of freely giving himself and entering into
communion with other persons."[37] Saint John Paul II stated that
the special love of the Creator for each human being "confers
upon him or her an infinite dignity."[38] Those who are committed to defending human dignity can find in the Christian faith
the deepest reasons for this commitment. How wonderful is
the certainty that each human life is not adrift in the midst of
hopeless chaos, in a world ruled by pure chance or endlessly
recurring cycles! The Creator can say to each one of us: "Before
I formed you in the womb, I knew you" (*Jer* 1:5). We were con-
ceived in the heart of God, and for this reason "each of us is the
result of a thought of God. Each of us is willed, each of us is
loved, each of us is necessary."[39]

66. The creation accounts in the book of Genesis contain, in
their own symbolic and narrative language, profound teachings

about human existence and its historical reality. They suggest that human life is grounded in three fundamental and closely intertwined relationships: with God, with our neighbour and with the earth itself. According to the Bible, these three vital relationships have been broken, both outwardly and within us. This rupture is sin. The harmony between the Creator, humanity and creation as a whole was disrupted by our presuming to take the place of God and refusing to acknowledge our creaturely limitations. This in turn distorted our mandate to "have dominion" over the earth (cf. *Gen* 1:28), to "till it and keep it" (*Gen* 2:15). As a result, the originally harmonious relationship between human beings and nature became conflictual (cf. *Gen* 3:17–19). It is significant that the harmony which Saint Francis of Assisi experienced with all creatures was seen as a healing of that rupture. Saint Bonaventure held that, through universal reconciliation with every creature, Saint Francis in some way returned to the state of original innocence.[40] This is a far cry from our situation today, where sin is manifest in all its destructive power in wars, the various forms of violence and abuse, the abandonment of the most vulnerable, and attacks on nature.

67. We are not God. The earth was here before us and it has been given to us. This allows us to respond to the charge that Judaeo-Christian thinking, on the basis of the Genesis account which grants man "dominion" over the earth (cf. *Gen* 1:28), has encouraged the unbridled exploitation of nature by painting him as domineering and destructive by nature. This is not a correct interpretation of the Bible as understood by the Church. Although it is true that we Christians have at times incorrectly

interpreted the Scriptures, nowadays we must forcefully reject the notion that our being created in God's image and given dominion over the earth justifies absolute domination over other creatures. The biblical texts are to be read in their context, with an appropriate hermeneutic, recognizing that they tell us to "till and keep" the garden of the world (cf. *Gen* 2:15). "Tilling" refers to cultivating, ploughing or working, while "keeping" means caring, protecting, overseeing and preserving. This implies a relationship of mutual responsibility between human beings and nature. Each community can take from the bounty of the earth whatever it needs for subsistence, but it also has the duty to protect the earth and to ensure its fruitfulness for coming generations. "The earth is the Lord's" (*Ps* 24:1); to him belongs "the earth with all that is within it" (*Dt* 10:14). Thus God rejects every claim to absolute ownership: "The land shall not be sold in perpetuity, for the land is mine; for you are strangers and sojourners with me" (*Lev* 25:23).

68. This responsibility for God's earth means that human beings, endowed with intelligence, must respect the laws of nature and the delicate equilibria existing between the creatures of this world, for "he commanded and they were created; and he established them for ever and ever; he fixed their bounds and he set a law which cannot pass away" (*Ps* 148:5b–6). The laws found in the Bible dwell on relationships, not only among individuals but also with other living beings. "You shall not see your brother's donkey or his ox fallen down by the way and withhold your help . . . If you chance to come upon a bird's nest in any tree or on the ground, with young ones or eggs and the mother sitting upon the young or upon the eggs; you shall

not take the mother with the young" (*Dt* 22:4, 6). Along these same lines, rest on the seventh day is meant not only for human beings, but also so "that your ox and your donkey may have rest" (*Ex* 23:12). Clearly, the Bible has no place for a tyrannical anthropocentrism unconcerned for other creatures.

69. Together with our obligation to use the earth's goods responsibly, we are called to recognize that other living beings have a value of their own in God's eyes: "by their mere existence they bless him and give him glory,"[41] and indeed, "the Lord rejoices in all his works" (*Ps* 104:31). By virtue of our unique dignity and our gift of intelligence, we are called to respect creation and its inherent laws, for "the Lord by wisdom founded the earth" (*Prov* 3:19). In our time, the Church does not simply state that other creatures are completely subordinated to the good of human beings, as if they have no worth in themselves and can be treated as we wish. The German bishops have taught that, where other creatures are concerned, "we can speak of the priority of *being* over that of *being useful*."[42] The Catechism clearly and forcefully criticizes a distorted anthropocentrism: "Each creature possesses its own particular goodness and perfection . . . Each of the various creatures, willed in its own being, reflects in its own way a ray of God's infinite wisdom and goodness. Man must therefore respect the particular goodness of every creature, to avoid any disordered use of things."[43]

70. In the story of Cain and Abel, we see how envy led Cain to commit the ultimate injustice against his brother, which in turn ruptured the relationship between Cain and God, and between Cain and the earth from which he was banished. This is

seen clearly in the dramatic exchange between God and Cain. God asks: "Where is Abel your brother?" Cain answers that he does not know, and God persists: "What have you done? The voice of your brother's blood is crying to me from the ground. And now you are cursed from the ground" (*Gen* 4:9–11). Disregard for the duty to cultivate and maintain a proper relationship with my neighbour, for whose care and custody I am responsible, ruins my relationship with my own self, with others, with God and with the earth. When all these relationships are neglected, when justice no longer dwells in the land, the Bible tells us that life itself is endangered. We see this in the story of Noah, where God threatens to do away with humanity because of its constant failure to fulfil the requirements of justice and peace: "I have determined to make an end of all flesh; for the earth is filled with violence through them" (*Gen* 6:13). These ancient stories, full of symbolism, bear witness to a conviction which we today share, that everything is interconnected, and that genuine care for our own lives and our relationships with nature is inseparable from fraternity, justice and faithfulness to others.

71. Although "the wickedness of man was great in the earth" (*Gen* 6:5) and the Lord "was sorry that he had made man on the earth" (*Gen* 6:6), nonetheless, through Noah, who remained innocent and just, God decided to open a path of salvation. In this way he gave humanity the chance of a new beginning. All it takes is one good person to restore hope! The biblical tradition clearly shows that this renewal entails recovering and respecting the rhythms inscribed in nature by the hand of the Creator. We see this, for example, in the law of the Sabbath.

On the seventh day, God rested from all his work. He commanded Israel to set aside each seventh day as a day of rest, a *Sabbath* (cf. *Gen* 2:2–3; *Ex* 16:23; 20:10). Similarly, every seven years, a sabbatical year was set aside for Israel, a complete rest for the land (cf. *Lev* 25:1–4), when sowing was forbidden and one reaped only what was necessary to live on and to feed one's household (cf. *Lev* 25:4–6). Finally, after seven weeks of years, which is to say forty-nine years, the Jubilee was celebrated as a year of general forgiveness and "liberty throughout the land for all its inhabitants" (cf. *Lev* 25:10). This law came about as an attempt to ensure balance and fairness in their relationships with others and with the land on which they lived and worked. At the same time, it was an acknowledgment that the gift of the earth with its fruits belongs to everyone. Those who tilled and kept the land were obliged to share its fruits, especially with the poor, with widows, orphans and foreigners in their midst: "When you reap the harvest of your land, you shall not reap your field to its very border, neither shall you gather the gleanings after the harvest. And you shall not strip your vineyard bare, neither shall you gather the fallen grapes of your vineyard; you shall leave them for the poor and for the sojourner" (*Lev* 19:9–10).

72. The Psalms frequently exhort us to praise God the Creator, "who spread out the earth on the waters, for his steadfast love endures for ever" (*Ps* 136:6). They also invite other creatures to join us in this praise: "Praise him, sun and moon, praise him, all you shining stars! Praise him, you highest heavens, and you waters above the heavens! Let them praise the name of the Lord, for he commanded and they were created" (*Ps* 148:3–5).

We do not only exist by God's mighty power; we also live with him and beside him. This is why we adore him.

73. The writings of the prophets invite us to find renewed strength in times of trial by contemplating the all-powerful God who created the universe. Yet God's infinite power does not lead us to flee his fatherly tenderness, because in him affection and strength are joined. Indeed, all sound spirituality entails both welcoming divine love and adoration, confident in the Lord because of his infinite power. In the Bible, the God who liberates and saves is the same God who created the universe, and these two divine ways of acting are intimately and inseparably connected: "Ah Lord God! It is you who made the heavens and the earth by your great power and by your outstretched arm! Nothing is too hard for you . . . You brought your people Israel out of the land of Egypt with signs and wonders" (*Jer* 32:17, 21). "The Lord is the everlasting God, the Creator of the ends of the earth. He does not faint or grow weary; his understanding is unsearchable. He gives power to the faint, and strengthens the powerless" (*Is* 40:28b–29).

74. The experience of the Babylonian captivity provoked a spiritual crisis which led to deeper faith in God. Now his creative omnipotence was given pride of place in order to exhort the people to regain their hope in the midst of their wretched predicament. Centuries later, in another age of trial and persecution, when the Roman Empire was seeking to impose absolute dominion, the faithful would once again find consolation and hope in a growing trust in the all-powerful God: "Great and wonderful are your deeds, O Lord God the

Almighty! Just and true are your ways!" (*Rev* 15:3). The God who created the universe out of nothing can also intervene in this world and overcome every form of evil. Injustice is not invincible.

75. A spirituality which forgets God as all-powerful and Creator is not acceptable. That is how we end up worshipping earthly powers, or ourselves usurping the place of God, even to the point of claiming an unlimited right to trample his creation underfoot. The best way to restore men and women to their rightful place, putting an end to their claim to absolute dominion over the earth, is to speak once more of the figure of a Father who creates and who alone owns the world. Otherwise, human beings will always try to impose their own laws and interests on reality.

III. The Mystery of the Universe

76. In the Judaeo-Christian tradition, the word "creation" has a broader meaning than "nature," for it has to do with God's loving plan in which every creature has its own value and significance. Nature is usually seen as a system which can be studied, understood and controlled, whereas creation can only be understood as a gift from the outstretched hand of the Father of all, and as a reality illuminated by the love which calls us together into universal communion.

77. "By the word of the Lord the heavens were made" (*Ps* 33:6). This tells us that the world came about as the result of a decision, not from chaos or chance, and this exalts it all the more. The creating word expresses a free choice. The universe did not emerge as the result of arbitrary omnipotence, a show of force or a desire for self-assertion. Creation is of the order of love. God's love is the fundamental moving force in all created things: "For you love all things that exist, and detest none of the things that you have made; for you would not have made anything if you had hated it" (*Wis* 11:24). Every creature is thus the object of the Father's tenderness, who gives it its place in the world. Even the fleeting life of the least of beings is the object of his love, and in its few seconds of existence, God enfolds it with his affection. Saint Basil the Great described the Creator as "goodness without measure,"[44] while Dante Alighieri spoke of "the love which moves the sun and the stars."[45] Consequently, we can ascend from created things "to the greatness of God and to his loving mercy."[46]

78. At the same time, Judaeo-Christian thought demythologized nature. While continuing to admire its grandeur and immensity, it no longer saw nature as divine. In doing so, it emphasizes all the more our human responsibility for nature. This rediscovery of nature can never be at the cost of the freedom and responsibility of human beings who, as part of the world, have the duty to cultivate their abilities in order to protect it and develop its potential. If we acknowledge the value and the fragility of nature and, at the same time, our God-given abilities, we can finally leave behind the modern myth of unlimited

material progress. A fragile world, entrusted by God to human care, challenges us to devise intelligent ways of directing, developing and limiting our power.

79. In this universe, shaped by open and intercommunicating systems, we can discern countless forms of relationship and participation. This leads us to think of the whole as open to God's transcendence, within which it develops. Faith allows us to interpret the meaning and the mysterious beauty of what is unfolding. We are free to apply our intelligence towards things evolving positively, or towards adding new ills, new causes of suffering and real setbacks. This is what makes for the excitement and drama of human history, in which freedom, growth, salvation and love can blossom, or lead towards decadence and mutual destruction. The work of the Church seeks not only to remind everyone of the duty to care for nature, but at the same time "she must above all protect mankind from self-destruction."[47]

80. Yet God, who wishes to work with us and who counts on our cooperation, can also bring good out of the evil we have done. "The Holy Spirit can be said to possess an infinite creativity, proper to the divine mind, which knows how to loosen the knots of human affairs, including the most complex and inscrutable."[48] Creating a world in need of development, God in some way sought to limit himself in such a way that many of the things we think of as evils, dangers or sources of suffering, are in reality part of the pains of childbirth which he uses to draw us into the act of cooperation with the Creator.[49] God is intimately present to each being, without impinging on the

autonomy of his creature, and this gives rise to the rightful autonomy of earthly affairs.[50] His divine presence, which ensures the subsistence and growth of each being, "continues the work of creation."[51] The Spirit of God has filled the universe with possibilities and therefore, from the very heart of things, something new can always emerge: "Nature is nothing other than a certain kind of art, namely God's art, impressed upon things, whereby those things are moved to a determinate end. It is as if a shipbuilder were able to give timbers the wherewithal to move themselves to take the form of a ship."[52]

81. Human beings, even if we postulate a process of evolution, also possess a uniqueness which cannot be fully explained by the evolution of other open systems. Each of us has his or her own personal identity and is capable of entering into dialogue with others and with God himself. Our capacity to reason, to develop arguments, to be inventive, to interpret reality and to create art, along with other not yet discovered capacities, are signs of a uniqueness which transcends the spheres of physics and biology. The sheer novelty involved in the emergence of a personal being within a material universe presupposes a direct action of God and a particular call to life and to relationship on the part of a "Thou" who addresses himself to another "thou." The biblical accounts of creation invite us to see each human being as a subject who can never be reduced to the status of an object.

82. Yet it would also be mistaken to view other living beings as mere objects subjected to arbitrary human domination. When nature is viewed solely as a source of profit and gain, this has serious consequences for society. This vision of "might is right"

has engendered immense inequality, injustice and acts of violence against the majority of humanity, since resources end up in the hands of the first comer or the most powerful: the winner takes all. Completely at odds with this model are the ideals of harmony, justice, fraternity and peace as proposed by Jesus. As he said of the powers of his own age: "You know that the rulers of the Gentiles lord it over them, and their great men exercise authority over them. It shall not be so among you; but whoever would be great among you must be your servant" (*Mt* 20:25–26).

83. The ultimate destiny of the universe is in the fullness of God, which has already been attained by the risen Christ, the measure of the maturity of all things.[53] Here we can add yet another argument for rejecting every tyrannical and irresponsible domination of human beings over other creatures. The ultimate purpose of other creatures is not to be found in us. Rather, all creatures are moving forward with us and through us towards a common point of arrival, which is God, in that transcendent fullness where the risen Christ embraces and illumines all things. Human beings, endowed with intelligence and love, and drawn by the fullness of Christ, are called to lead all creatures back to their Creator.

IV. The Message of Each Creature in the Harmony of Creation

84. Our insistence that each human being is an image of God should not make us overlook the fact that each creature has its

own purpose. None is superfluous. The entire material universe speaks of God's love, his boundless affection for us. Soil, water, mountains: everything is, as it were, a caress of God. The history of our friendship with God is always linked to particular places which take on an intensely personal meaning; we all remember places, and revisiting those memories does us much good. Anyone who has grown up in the hills or used to sit by the spring to drink, or played outdoors in the neighbourhood square; going back to these places is a chance to recover something of their true selves.

85. God has written a precious book, "whose letters are the multitude of created things present in the universe."[54] The Canadian bishops rightly pointed out that no creature is excluded from this manifestation of God: "From panoramic vistas to the tiniest living form, nature is a constant source of wonder and awe. It is also a continuing revelation of the divine."[55] The bishops of Japan, for their part, made a thought-provoking observation: "To sense each creature singing the hymn of its existence is to live joyfully in God's love and hope."[56] This contemplation of creation allows us to discover in each thing a teaching which God wishes to hand on to us, since "for the believer, to contemplate creation is to hear a message, to listen to a paradoxical and silent voice."[57] We can say that "alongside revelation properly so-called, contained in sacred Scripture, there is a divine manifestation in the blaze of the sun and the fall of night."[58] Paying attention to this manifestation, we learn to see ourselves in relation to all other creatures: "I express myself in expressing the world; in my effort to decipher the sacredness of the world, I explore my own."[59]

86. The universe as a whole, in all its manifold relationships, shows forth the inexhaustible riches of God. Saint Thomas Aquinas wisely noted that multiplicity and variety "come from the intention of the first agent" who willed that "what was wanting to one in the representation of the divine goodness might be supplied by another,"[60] inasmuch as God's goodness "could not be represented fittingly by any one creature."[61] Hence we need to grasp the variety of things in their multiple relationships.[62] We understand better the importance and meaning of each creature if we contemplate it within the entirety of God's plan. As the Catechism teaches: "God wills the interdependence of creatures. The sun and the moon, the cedar and the little flower, the eagle and the sparrow: the spectacle of their countless diversities and inequalities tells us that no creature is self-sufficient. Creatures exist only in dependence on each other, to complete each other, in the service of each other."[63]

87. When we can see God reflected in all that exists, our hearts are moved to praise the Lord for all his creatures and to worship him in union with them. This sentiment finds magnificent expression in the hymn of Saint Francis of Assisi:

Praised be you, my Lord, with all your creatures,
especially Sir Brother Sun,
who is the day and through whom you give us light.
And he is beautiful and radiant with great splendour;
and bears a likeness of you, Most High.
Praised be you, my Lord, through Sister Moon and the stars,
in heaven you formed them clear and precious and beautiful.
Praised be you, my Lord, through Brother Wind,

and through the air, cloudy and serene, and every kind of weather
through whom you give sustenance to your creatures.
Praised be you, my Lord, through Sister Water,
who is very useful and humble and precious and chaste.
Praised be you, my Lord, through Brother Fire,
through whom you light the night,
and he is beautiful and playful and robust and strong.[64]

88. The bishops of Brazil have pointed out that nature as a whole not only manifests God but is also a locus of his presence. The Spirit of life dwells in every living creature and calls us to enter into relationship with him.[65] Discovering this presence leads us to cultivate the "ecological virtues."[66] This is not to forget that there is an infinite distance between God and the things of this world, which do not possess his fullness. Otherwise, we would not be doing the creatures themselves any good either, for we would be failing to acknowledge their right and proper place. We would end up unduly demanding of them something which they, in their smallness, cannot give us.

V. A Universal Communion

89. The created things of this world are not free of ownership: "For they are yours, O Lord, who love the living" (*Wis* 11:26). This is the basis of our conviction that, as part of the universe, called into being by one Father, all of us are linked by unseen bonds and together form a kind of universal family, a sublime communion which fills us with a sacred, affectionate and

humble respect. Here I would reiterate that "God has joined us so closely to the world around us that we can feel the desertification of the soil almost as a physical ailment, and the extinction of a species as a painful disfigurement."[67]

90. This is not to put all living beings on the same level nor to deprive human beings of their unique worth and the tremendous responsibility it entails. Nor does it imply a divinization of the earth which would prevent us from working on it and protecting it in its fragility. Such notions would end up creating new imbalances which would deflect us from the reality which challenges us.[68] At times we see an obsession with denying any pre-eminence to the human person; more zeal is shown in protecting other species than in defending the dignity which all human beings share in equal measure. Certainly, we should be concerned lest other living beings be treated irresponsibly. But we should be particularly indignant at the enormous inequalities in our midst, whereby we continue to tolerate some considering themselves more worthy than others. We fail to see that some are mired in desperate and degrading poverty, with no way out, while others have not the faintest idea of what to do with their possessions, vainly showing off their supposed superiority and leaving behind them so much waste which, if it were the case everywhere, would destroy the planet. In practice, we continue to tolerate that some consider themselves more human than others, as if they had been born with greater rights.

91. A sense of deep communion with the rest of nature cannot be real if our hearts lack tenderness, compassion and concern for our fellow human beings. It is clearly inconsistent to combat

trafficking in endangered species while remaining completely indifferent to human trafficking, unconcerned about the poor, or undertaking to destroy another human being deemed unwanted. This compromises the very meaning of our struggle for the sake of the environment. It is no coincidence that, in the canticle in which Saint Francis praises God for his creatures, he goes on to say: "Praised be you my Lord, through those who give pardon for your love." Everything is connected. Concern for the environment thus needs to be joined to a sincere love for our fellow human beings and an unwavering commitment to resolving the problems of society.

92. Moreover, when our hearts are authentically open to universal communion, this sense of fraternity excludes nothing and no one. It follows that our indifference or cruelty towards fellow creatures of this world sooner or later affects the treatment we mete out to other human beings. We have only one heart, and the same wretchedness which leads us to mistreat an animal will not be long in showing itself in our relationships with other people. Every act of cruelty towards any creature is "contrary to human dignity."[69] We can hardly consider ourselves to be fully loving if we disregard any aspect of reality: "Peace, justice and the preservation of creation are three absolutely interconnected themes, which cannot be separated and treated individually without once again falling into reductionism."[70] Everything is related, and we human beings are united as brothers and sisters on a wonderful pilgrimage, woven together by the love God has for each of his creatures and which also unites us in fond affection with brother sun, sister moon, brother river and mother earth.

VI. The Common Destination
of Goods

93. Whether believers or not, we are agreed today that the earth is essentially a shared inheritance, whose fruits are meant to benefit everyone. For believers, this becomes a question of fidelity to the Creator, since God created the world for everyone. Hence every ecological approach needs to incorporate a social perspective which takes into account the fundamental rights of the poor and the underprivileged. The principle of the subordination of private property to the universal destination of goods, and thus the right of everyone to their use, is a golden rule of social conduct and "the first principle of the whole ethical and social order."[71] The Christian tradition has never recognized the right to private property as absolute or inviolable, and has stressed the social purpose of all forms of private property. Saint John Paul II forcefully reaffirmed this teaching, stating that "God gave the earth to the whole human race for the sustenance of all its members, *without excluding or favouring anyone*."[72] These are strong words. He noted that "a type of development which did not respect and promote human rights—personal and social, economic and political, including the rights of nations and of peoples—would not be really worthy of man."[73] He clearly explained that "the Church does indeed defend the legitimate right to private property, but she also teaches no less clearly that there is always a social mortgage on all private property, in order that goods may serve the general purpose that God gave them."[74] Consequently,

he maintained, "it is not in accord with God's plan that this gift be used in such a way that its benefits favour only a few."[75] This calls into serious question the unjust habits of a part of humanity.[76]

94. The rich and the poor have equal dignity, for "the Lord is the maker of them all" (*Prov* 22:2). "He himself made both small and great" (*Wis* 6:7), and "he makes his sun rise on the evil and on the good" (*Mt* 5:45). This has practical consequences, such as those pointed out by the bishops of Paraguay: "Every *campesino* has a natural right to possess a reasonable allotment of land where he can establish his home, work for subsistence of his family and a secure life. This right must be guaranteed so that its exercise is not illusory but real. That means that apart from the ownership of property, rural people must have access to means of technical education, credit, insurance, and markets."[77]

95. The natural environment is a collective good, the patrimony of all humanity and the responsibility of everyone. If we make something our own, it is only to administer it for the good of all. If we do not, we burden our consciences with the weight of having denied the existence of others. That is why the New Zealand bishops asked what the commandment "Thou shall not kill" means when "twenty percent of the world's population consumes resources at a rate that robs the poor nations and future generations of what they need to survive."[78]

VII. The Gaze of Jesus

96. Jesus took up the biblical faith in God the Creator, emphasizing a fundamental truth: God is Father (cf. *Mt* 11:25). In talking with his disciples, Jesus would invite them to recognize the paternal relationship God has with all his creatures. With moving tenderness he would remind them that each one of them is important in God's eyes: "Are not five sparrows sold for two pennies? And not one of them is forgotten before God" (*Lk* 12:6). "Look at the birds of the air: they neither sow nor reap nor gather into barns, and yet your heavenly Father feeds them" (*Mt* 6:26).

97. The Lord was able to invite others to be attentive to the beauty that there is in the world because he himself was in constant touch with nature, lending it an attention full of fondness and wonder. As he made his way throughout the land, he often stopped to contemplate the beauty sown by his Father, and invited his disciples to perceive a divine message in things: "Lift up your eyes, and see how the fields are already white for harvest" (*Jn* 4:35). "The kingdom of God is like a grain of mustard seed which a man took and sowed in his field; it is the smallest of all seeds, but once it has grown, it is the greatest of plants" (*Mt* 13:31–32).

98. Jesus lived in full harmony with creation, and others were amazed: "What sort of man is this, that even the winds and the sea obey him?" (*Mt* 8:27). His appearance was not that of an ascetic set apart from the world, nor of an enemy to the

pleasant things of life. Of himself he said: "The Son of Man came eating and drinking and they say, 'Look, a glutton and a drunkard!'" (*Mt* 11:19). He was far removed from philosophies which despised the body, matter and the things of the world. Such unhealthy dualisms, nonetheless, left a mark on certain Christian thinkers in the course of history and disfigured the Gospel. Jesus worked with his hands, in daily contact with the matter created by God, to which he gave form by his craftsmanship. It is striking that most of his life was dedicated to this task in a simple life which awakened no admiration at all: "Is not this the carpenter, the son of Mary?" (*Mk* 6:3). In this way he sanctified human labour and endowed it with a special significance for our development. As Saint John Paul II taught, "by enduring the toil of work in union with Christ crucified for us, man in a way collaborates with the Son of God for the redemption of humanity."[79]

99. In the Christian understanding of the world, the destiny of all creation is bound up with the mystery of Christ, present from the beginning: "All things have been created though him and for him" (*Col* 1:16).[80] The prologue of the Gospel of John (1:1–18) reveals Christ's creative work as the Divine Word (*Logos*). But then, unexpectedly, the prologue goes on to say that this same Word "became flesh" (*Jn* 1:14). One Person of the Trinity entered into the created cosmos, throwing in his lot with it, even to the cross. From the beginning of the world, but particularly through the incarnation, the mystery of Christ is at work in a hidden manner in the natural world as a whole, without thereby impinging on its autonomy.

100. The New Testament does not only tell us of the earthly Jesus and his tangible and loving relationship with the world. It also shows him risen and glorious, present throughout creation by his universal Lordship: "For in him all the fullness of God was pleased to dwell, and through him to reconcile to himself all things, whether on earth or in heaven, making peace by the blood of his cross" (*Col* 1:19–20). This leads us to direct our gaze to the end of time, when the Son will deliver all things to the Father, so that "God may be everything to every one" (*1 Cor* 15:28). Thus, the creatures of this world no longer appear to us under merely natural guise because the risen One is mysteriously holding them to himself and directing them towards fullness as their end. The very flowers of the field and the birds which his human eyes contemplated and admired are now imbued with his radiant presence.

CHAPTER THREE

The Human Roots of the Ecological Crisis

101. It would hardly be helpful to describe symptoms without acknowledging the human origins of the ecological crisis. A certain way of understanding human life and activity has gone awry, to the serious detriment of the world around us. Should we not pause and consider this? At this stage, I propose that we focus on the dominant technocratic paradigm and the place of human beings and of human action in the world.

I. Technology: Creativity and Power

102. Humanity has entered a new era in which our technical prowess has brought us to a crossroads. We are the beneficiaries of two centuries of enormous waves of change: steam engines,

railways, the telegraph, electricity, automobiles, aeroplanes, chemical industries, modern medicine, information technology and, more recently, the digital revolution, robotics, biotechnologies and nanotechnologies. It is right to rejoice in these advances and to be excited by the immense possibilities which they continue to open up before us, for "science and technology are wonderful products of a God-given human creativity."[81] The modification of nature for useful purposes has distinguished the human family from the beginning; technology itself "expresses the inner tension that impels man gradually to overcome material limitations."[82] Technology has remedied countless evils which used to harm and limit human beings. How can we not feel gratitude and appreciation for this progress, especially in the fields of medicine, engineering and communications? How could we not acknowledge the work of many scientists and engineers who have provided alternatives to make development sustainable?

103. Technoscience, when well directed, can produce important means of improving the quality of human life, from useful domestic appliances to great transportation systems, bridges, buildings and public spaces. It can also produce art and enable men and women immersed in the material world to "leap" into the world of beauty. Who can deny the beauty of an aircraft or a skyscraper? Valuable works of art and music now make use of new technologies. So, in the beauty intended by the one who uses new technical instruments and in the contemplation of such beauty, a quantum leap occurs, resulting in a fulfilment which is uniquely human.

104. Yet it must also be recognized that nuclear energy, bio-technology, information technology, knowledge of our DNA, and many other abilities which we have acquired, have given us tremendous power. More precisely, they have given those with the knowledge, and especially the economic resources to use them, an impressive dominance over the whole of humanity and the entire world. Never has humanity had such power over itself, yet nothing ensures that it will be used wisely, particularly when we consider how it is currently being used. We need but think of the nuclear bombs dropped in the middle of the twentieth century, or the array of technology which Nazism, Communism and other totalitarian regimes have employed to kill millions of people, to say nothing of the increasingly deadly arsenal of weapons available for modern warfare. In whose hands does all this power lie, or will it eventually end up? It is extremely risky for a small part of humanity to have it.

105. There is a tendency to believe that every increase in power means "an increase of 'progress' itself," an advance in "security, usefulness, welfare and vigour; . . . an assimilation of new values into the stream of culture,"[83] as if reality, goodness and truth automatically flow from technological and economic power as such. The fact is that "contemporary man has not been trained to use power well,"[84] because our immense technological development has not been accompanied by a development in human responsibility, values and conscience. Each age tends to have only a meagre awareness of its own limitations. It is possible that we do not grasp the gravity of the challenges now before us. "The risk is growing day by day that man will not use

his power as he should"; in effect, "power is never considered in terms of the responsibility of choice which is inherent in freedom" since its "only norms are taken from alleged necessity, from either utility or security."[85] But human beings are not completely autonomous. Our freedom fades when it is handed over to the blind forces of the unconscious, of immediate needs, of self-interest, and of violence. In this sense, we stand naked and exposed in the face of our ever-increasing power, lacking the wherewithal to control it. We have certain superficial mechanisms, but we cannot claim to have a sound ethics, a culture and spirituality genuinely capable of setting limits and teaching clear-minded self-restraint.

II. The Globalization of the Technocratic Paradigm

106. The basic problem goes even deeper: it is the way that humanity has taken up technology and its development *according to an undifferentiated and one-dimensional paradigm*. This paradigm exalts the concept of a subject who, using logical and rational procedures, progressively approaches and gains control over an external object. This subject makes every effort to establish the scientific and experimental method, which in itself is already a technique of possession, mastery and transformation. It is as if the subject were to find itself in the presence of something formless, completely open to manipulation. Men and women have constantly intervened in nature, but for a long time this meant being in tune with and respecting the possibilities

offered by the things themselves. It was a matter of receiving what nature itself allowed, as if from its own hand. Now, by contrast, we are the ones to lay our hands on things, attempting to extract everything possible from them while frequently ignoring or forgetting the reality in front of us. Human beings and material objects no longer extend a friendly hand to one another; the relationship has become confrontational. This has made it easy to accept the idea of infinite or unlimited growth, which proves so attractive to economists, financiers and experts in technology. It is based on the lie that there is an infinite supply of the earth's goods, and this leads to the planet being squeezed dry beyond every limit. It is the false notion that "an infinite quantity of energy and resources are available, that it is possible to renew them quickly, and that the negative effects of the exploitation of the natural order can be easily absorbed."[86]

107. It can be said that many problems of today's world stem from the tendency, at times unconscious, to make the method and aims of science and technology an epistemological paradigm which shapes the lives of individuals and the workings of society. The effects of imposing this model on reality as a whole, human and social, are seen in the deterioration of the environment, but this is just one sign of a reductionism which affects every aspect of human and social life. We have to accept that technological products are not neutral, for they create a framework which ends up conditioning lifestyles and shaping social possibilities along the lines dictated by the interests of certain powerful groups. Decisions which may seem purely instrumental are in reality decisions about the kind of society we want to build.

108. The idea of promoting a different cultural paradigm and employing technology as a mere instrument is nowadays inconceivable. The technological paradigm has become so dominant that it would be difficult to do without its resources and even more difficult to utilize them without being dominated by their internal logic. It has become countercultural to choose a lifestyle whose goals are even partly independent of technology, of its costs and its power to globalize and make us all the same. Technology tends to absorb everything into its ironclad logic, and those who are surrounded with technology "know full well that it moves forward in the final analysis neither for profit nor for the well-being of the human race," that "in the most radical sense of the term power is its motive—a lordship over all."[87] As a result, "man seizes hold of the naked elements of both nature and human nature."[88] Our capacity to make decisions, a more genuine freedom and the space for each one's alternative creativity are diminished.

109. The technocratic paradigm also tends to dominate economic and political life. The economy accepts every advance in technology with a view to profit, without concern for its potentially negative impact on human beings. Finance overwhelms the real economy. The lessons of the global financial crisis have not been assimilated, and we are learning all too slowly the lessons of environmental deterioration. Some circles maintain that current economics and technology will solve all environmental problems, and argue, in popular and non-technical terms, that the problems of global hunger and poverty will be resolved simply by market growth. They are less concerned with certain economic theories which today scarcely anybody dares defend,

than with their actual operation in the functioning of the economy. They may not affirm such theories with words, but nonetheless support them with their deeds by showing no interest in more balanced levels of production, a better distribution of wealth, concern for the environment and the rights of future generations. Their behaviour shows that for them maximizing profits is enough. Yet by itself the market cannot guarantee integral human development and social inclusion.[89] At the same time, we have "a sort of 'superdevelopment' of a wasteful and consumerist kind which forms an unacceptable contrast with the ongoing situations of dehumanizing deprivation,"[90] while we are all too slow in developing economic institutions and social initiatives which can give the poor regular access to basic resources. We fail to see the deepest roots of our present failures, which have to do with the direction, goals, meaning and social implications of technological and economic growth.

110. The specialization which belongs to technology makes it difficult to see the larger picture. The fragmentation of knowledge proves helpful for concrete applications, and yet it often leads to a loss of appreciation for the whole, for the relationships between things, and for the broader horizon, which then becomes irrelevant. This very fact makes it hard to find adequate ways of solving the more complex problems of today's world, particularly those regarding the environment and the poor; these problems cannot be dealt with from a single perspective or from a single set of interests. A science which would offer solutions to the great issues would necessarily have to take into account the data generated by other fields of knowledge, including philosophy and social ethics; but this is a difficult

habit to acquire today. Nor are there genuine ethical horizons to which one can appeal. Life gradually becomes a surrender to situations conditioned by technology, itself viewed as the principal key to the meaning of existence. In the concrete situation confronting us, there are a number of symptoms which point to what is wrong, such as environmental degradation, anxiety, a loss of the purpose of life and of community living. Once more we see that "realities are more important than ideas."[91]

111. Ecological culture cannot be reduced to a series of urgent and partial responses to the immediate problems of pollution, environmental decay and the depletion of natural resources. There needs to be a distinctive way of looking at things, a way of thinking, policies, an educational programme, a lifestyle and a spirituality which together generate resistance to the assault of the technocratic paradigm. Otherwise, even the best ecological initiatives can find themselves caught up in the same globalized logic. To seek only a technical remedy to each environmental problem which comes up is to separate what is in reality interconnected and to mask the true and deepest problems of the global system.

112. Yet we can once more broaden our vision. We have the freedom needed to limit and direct technology; we can put it at the service of another type of progress, one which is healthier, more human, more social, more integral. Liberation from the dominant technocratic paradigm does in fact happen sometimes, for example, when cooperatives of small producers adopt less polluting means of production, and opt for a non-consumerist

model of life, recreation and community. Or when technology is directed primarily to resolving people's concrete problems, truly helping them live with more dignity and less suffering. Or indeed when the desire to create and contemplate beauty manages to overcome reductionism through a kind of salvation which occurs in beauty and in those who behold it. An authentic humanity, calling for a new synthesis, seems to dwell in the midst of our technological culture, almost unnoticed, like a mist seeping gently beneath a closed door. Will the promise last, in spite of everything, with all that is authentic rising up in stubborn resistance?

113. There is also the fact that people no longer seem to believe in a happy future; they no longer have blind trust in a better tomorrow based on the present state of the world and our technical abilities. There is a growing awareness that scientific and technological progress cannot be equated with the progress of humanity and history, a growing sense that the way to a better future lies elsewhere. This is not to reject the possibilities which technology continues to offer us. But humanity has changed profoundly, and the accumulation of constant novelties exalts a superficiality which pulls us in one direction. It becomes difficult to pause and recover depth in life. If architecture reflects the spirit of an age, our megastructures and drab apartment blocks express the spirit of globalized technology, where a constant flood of new products coexists with a tedious monotony. Let us refuse to resign ourselves to this, and continue to wonder about the purpose and meaning of everything. Otherwise we would simply legitimate the present situation and need new forms of escapism to help us endure the emptiness.

114. All of this shows the urgent need for us to move forward in a bold cultural revolution. Science and technology are not neutral; from the beginning to the end of a process, various intentions and possibilities are in play and can take on distinct shapes. Nobody is suggesting a return to the Stone Age, but we do need to slow down and look at reality in a different way, to appropriate the positive and sustainable progress which has been made, but also to recover the values and the great goals swept away by our unrestrained delusions of grandeur.

III. The Crisis and Effects of Modern Anthropocentrism

115. Modern anthropocentrism has paradoxically ended up prizing technical thought over reality, since "the technological mind sees nature as an insensate order, as a cold body of facts, as a mere 'given,' as an object of utility, as raw material to be hammered into useful shape; it views the cosmos similarly as a mere 'space' into which objects can be thrown with complete indifference."[92] The intrinsic dignity of the world is thus compromised. When human beings fail to find their true place in this world, they misunderstand themselves and end up acting against themselves: "Not only has God given the earth to man, who must use it with respect for the original good purpose for which it was given, but, man too is God's gift to man. He must therefore respect the natural and moral structure with which he has been endowed."[93]

116. Modernity has been marked by an excessive anthropocentrism which today, under another guise, continues to stand in the way of shared understanding and of any effort to strengthen social bonds. The time has come to pay renewed attention to reality and the limits it imposes; this in turn is the condition for a more sound and fruitful development of individuals and society. An inadequate presentation of Christian anthropology gave rise to a wrong understanding of the relationship between human beings and the world. Often, what was handed on was a Promethean vision of mastery over the world, which gave the impression that the protection of nature was something that only the faint-hearted cared about. Instead, our "dominion" over the universe should be understood more properly in the sense of responsible stewardship.[94]

117. Neglecting to monitor the harm done to nature and the environmental impact of our decisions is only the most striking sign of a disregard for the message contained in the structures of nature itself. When we fail to acknowledge as part of reality the worth of a poor person, a human embryo, a person with disabilities—to offer just a few examples—it becomes difficult to hear the cry of nature itself; everything is connected. Once the human being declares independence from reality and behaves with absolute dominion, the very foundations of our life begin to crumble, for "instead of carrying out his role as a cooperator with God in the work of creation, man sets himself up in place of God and thus ends up provoking a rebellion on the part of nature."[95]

118. This situation has led to a constant schizophrenia, wherein a technocracy which sees no intrinsic value in lesser beings

coexists with the other extreme, which sees no special value in human beings. But one cannot prescind from humanity. There can be no renewal of our relationship with nature without a renewal of humanity itself. There can be no ecology without an adequate anthropology. When the human person is considered as simply one being among others, the product of chance or physical determinism, then "our overall sense of responsibility wanes."[96] A misguided anthropocentrism need not necessarily yield to "biocentrism," for that would entail adding yet another imbalance, failing to solve present problems and adding new ones. Human beings cannot be expected to feel responsibility for the world unless, at the same time, their unique capacities of knowledge, will, freedom and responsibility are recognized and valued.

119. Nor must the critique of a misguided anthropocentrism underestimate the importance of interpersonal relations. If the present ecological crisis is one small sign of the ethical, cultural and spiritual crisis of modernity, we cannot presume to heal our relationship with nature and the environment without healing all fundamental human relationships. Christian thought sees human beings as possessing a particular dignity above other creatures; it thus inculcates esteem for each person and respect for others. Our openness to others, each of whom is a "thou" capable of knowing, loving and entering into dialogue, remains the source of our nobility as human persons. A correct relationship with the created world demands that we not weaken this social dimension of openness to others, much less the transcendent dimension of our openness to the "Thou" of God. Our relationship with the environment can never be isolated from our

relationship with others and with God. Otherwise, it would be nothing more than romantic individualism dressed up in ecological garb, locking us into a stifling immanence.

120. Since everything is interrelated, concern for the protection of nature is also incompatible with the justification of abortion. How can we genuinely teach the importance of concern for other vulnerable beings, however troublesome or inconvenient they may be, if we fail to protect a human embryo, even when its presence is uncomfortable and creates difficulties? "If personal and social sensitivity towards the acceptance of the new life is lost, then other forms of acceptance that are valuable for society also wither away."[97]

121. We need to develop a new synthesis capable of overcoming the false arguments of recent centuries. Christianity, in fidelity to its own identity and the rich deposit of truth which it has received from Jesus Christ, continues to reflect on these issues in fruitful dialogue with changing historical situations. In doing so, it reveals its eternal newness.[98]

PRACTICAL RELATIVISM

122. A misguided anthropocentrism leads to a misguided lifestyle. In the Apostolic Exhortation *Evangelii Gaudium*, I noted that the practical relativism typical of our age is "even more dangerous than doctrinal relativism."[99] When human beings place themselves at the centre, they give absolute priority to immediate convenience and all else becomes relative. Hence

we should not be surprised to find, in conjunction with the omnipresent technocratic paradigm and the cult of unlimited human power, the rise of a relativism which sees everything as irrelevant unless it serves one's own immediate interests. There is a logic in all this whereby different attitudes can feed on one another, leading to environmental degradation and social decay.

123. The culture of relativism is the same disorder which drives one person to take advantage of another, to treat others as mere objects, imposing forced labour on them or enslaving them to pay their debts. The same kind of thinking leads to the sexual exploitation of children and abandonment of the elderly who no longer serve our interests. It is also the mindset of those who say: Let us allow the invisible forces of the market to regulate the economy, and consider their impact on society and nature as collateral damage. In the absence of objective truths or sound principles other than the satisfaction of our own desires and immediate needs, what limits can be placed on human trafficking, organized crime, the drug trade, commerce in blood diamonds and the fur of endangered species? Is it not the same relativistic logic which justifies buying the organs of the poor for resale or use in experimentation, or eliminating children because they are not what their parents wanted? This same "use and throw away" logic generates so much waste, because of the disordered desire to consume more than what is really necessary. We should not think that political efforts or the force of law will be sufficient to prevent actions which affect the environment because, when the culture itself is corrupt and objective truth and universally valid principles are no longer upheld,

then laws can only be seen as arbitrary impositions or obstacles to be avoided.

THE NEED TO PROTECT EMPLOYMENT

124. Any approach to an integral ecology, which by definition does not exclude human beings, needs to take account of the value of labour, as Saint John Paul II wisely noted in his Encyclical *Laborem Exercens*. According to the biblical account of creation, God placed man and woman in the garden he had created (cf. *Gen* 2:15) not only to preserve it ("keep") but also to make it fruitful ("till"). Labourers and craftsmen thus "maintain the fabric of the world" (*Sir* 38:34). Developing the created world in a prudent way is the best way of caring for it, as this means that we ourselves become the instrument used by God to bring out the potential which he himself inscribed in things: "The Lord created medicines out of the earth, and a sensible man will not despise them" (*Sir* 38:4).

125. If we reflect on the proper relationship between human beings and the world around us, we see the need for a correct understanding of work; if we talk about the relationship between human beings and things, the question arises as to the meaning and purpose of all human activity. This has to do not only with manual or agricultural labour but with any activity involving a modification of existing reality, from producing a social report to the design of a technological development. Underlying every form of work is a concept of the relationship which we can and must have with what is other than ourselves. Together with

the awe-filled contemplation of creation which we find in Saint
Francis of Assisi, the Christian spiritual tradition has also developed a rich and balanced understanding of the meaning of
work, as, for example, in the life of Blessed Charles de Fou-
cauld and his followers.

126. We can also look to the great tradition of monasticism.
Originally, it was a kind of flight from the world, an escape
from the decadence of the cities. The monks sought the des-
ert, convinced that it was the best place for encountering the
presence of God. Later, Saint Benedict of Norcia proposed that
his monks live in community, combining prayer and spiritual
reading with manual labour (*ora et labora*). Seeing manual labour as spiritually meaningful proved revolutionary. Personal
growth and sanctification came to be sought in the interplay of
recollection and work. This way of experiencing work makes us
more protective and respectful of the environment; it imbues
our relationship to the world with a healthy sobriety.

127. We are convinced that "man is the source, the focus and the
aim of all economic and social life."[100] Nonetheless, once our
human capacity for contemplation and reverence is impaired, it
becomes easy for the meaning of work to be misunderstood.[101]
We need to remember that men and women have "the capacity to improve their lot, to further their moral growth and to
develop their spiritual endowments."[102] Work should be the setting for this rich personal growth, where many aspects of life
enter into play: creativity, planning for the future, developing
our talents, living out our values, relating to others, giving glory
to God. It follows that, in the reality of today's global society,

it is essential that "we continue to prioritize the goal of access to steady employment for everyone,"[103] no matter the limited interests of business and dubious economic reasoning.

128. We were created with a vocation to work. The goal should not be that technological progress increasingly replace human work, for this would be detrimental to humanity. Work is a necessity, part of the meaning of life on this earth, a path to growth, human development and personal fulfilment. Helping the poor financially must always be a provisional solution in the face of pressing needs. The broader objective should always be to allow them a dignified life through work. Yet the orientation of the economy has favoured a kind of technological progress in which the costs of production are reduced by laying off workers and replacing them with machines. This is yet another way in which we can end up working against ourselves. The loss of jobs also has a negative impact on the economy "through the progressive erosion of social capital: the network of relationships of trust, dependability, and respect for rules, all of which are indispensable for any form of civil coexistence."[104] In other words, "human costs always include economic costs, and economic dysfunctions always involve human costs."[105] To stop investing in people, in order to gain greater short-term financial gain, is bad business for society.

129. In order to continue providing employment, it is imperative to promote an economy which favours productive diversity and business creativity. For example, there is a great variety of small-scale food production systems which feed the greater part of the world's peoples, using a modest amount of land

and producing less waste, be it in small agricultural parcels, in orchards and gardens, hunting and wild harvesting or local fishing. Economies of scale, especially in the agricultural sector, end up forcing smallholders to sell their land or to abandon their traditional crops. Their attempts to move to other, more diversified, means of production prove fruitless because of the difficulty of linkage with regional and global markets, or because the infrastructure for sales and transport is geared to larger businesses. Civil authorities have the right and duty to adopt clear and firm measures in support of small producers and differentiated production. To ensure economic freedom from which all can effectively benefit, restraints occasionally have to be imposed on those possessing greater resources and financial power. To claim economic freedom while real conditions bar many people from actual access to it, and while possibilities for employment continue to shrink, is to practise a doublespeak which brings politics into disrepute. Business is a noble vocation, directed to producing wealth and improving our world. It can be a fruitful source of prosperity for the areas in which it operates, especially if it sees the creation of jobs as an essential part of its service to the common good.

NEW BIOLOGICAL TECHNOLOGIES

130. In the philosophical and theological vision of the human being and of creation which I have presented, it is clear that the human person, endowed with reason and knowledge, is not an external factor to be excluded. While human intervention

on plants and animals is permissible when it pertains to the necessities of human life, the *Catechism of the Catholic Church* teaches that experimentation on animals is morally acceptable only "if it remains within reasonable limits [and] contributes to caring for or saving human lives."[106] The *Catechism* firmly states that human power has limits and that "it is contrary to human dignity to cause animals to suffer or die needlessly."[107] All such use and experimentation "requires a religious respect for the integrity of creation."[108]

131. Here I would recall the balanced position of Saint John Paul II, who stressed the benefits of scientific and technological progress as evidence of "the nobility of the human vocation to participate responsibly in God's creative action," while also noting that "we cannot interfere in one area of the ecosystem without paying due attention to the consequences of such interference in other areas."[109] He made it clear that the Church values the benefits which result "from the study and applications of molecular biology, supplemented by other disciplines such as genetics, and its technological application in agriculture and industry."[110] But he also pointed out that this should not lead to "indiscriminate genetic manipulation"[111] which ignores the negative effects of such interventions. Human creativity cannot be suppressed. If an artist cannot be stopped from using his or her creativity, neither should those who possess particular gifts for the advancement of science and technology be prevented from using their God-given talents for the service of others. We need constantly to rethink the goals, effects, overall context and ethical limits of this human activity, which is a form of power involving considerable risks.

132. This, then, is the correct framework for any reflection concerning human intervention on plants and animals, which at present includes genetic manipulation by biotechnology for the sake of exploiting the potential present in material reality. The respect owed by faith to reason calls for close attention to what the biological sciences, through research uninfluenced by economic interests, can teach us about biological structures, their possibilities and their mutations. Any legitimate intervention will act on nature only in order "to favour its development in its own line, that of creation, as intended by God."[112]

133. It is difficult to make a general judgement about genetic modification (GM), whether vegetable or animal, medical or agricultural, since these vary greatly among themselves and call for specific considerations. The risks involved are not always due to the techniques used, but rather to their improper or excessive application. Genetic mutations, in fact, have often been, and continue to be, caused by nature itself. Nor are mutations caused by human intervention a modern phenomenon. The domestication of animals, the crossbreeding of species and other older and universally accepted practices can be mentioned as examples. We need but recall that scientific developments in GM cereals began with the observation of natural bacteria which spontaneously modified plant genomes. In nature, however, this process is slow and cannot be compared to the fast pace induced by contemporary technological advances, even when the latter build upon several centuries of scientific progress.

134. Although no conclusive proof exists that GM cereals may be harmful to human beings, and in some regions their use

has brought about economic growth which has helped to resolve problems, there remain a number of significant difficulties
which should not be underestimated. In many places, following the introduction of these crops, productive land is concentrated in the hands of a few owners due to "the progressive
disappearance of small producers, who, as a consequence of the
loss of the exploited lands, are obliged to withdraw from direct
production."[113] The most vulnerable of these become temporary
labourers, and many rural workers end up moving to poverty-
stricken urban areas. The expansion of these crops has the
effect of destroying the complex network of ecosystems, diminishing the diversity of production and affecting regional
economies, now and in the future. In various countries, we see
an expansion of oligopolies for the production of cereals and
other products needed for their cultivation. This dependency
would be aggravated were the production of infertile seeds to
be considered; the effect would be to force farmers to purchase
them from larger producers.

135. Certainly, these issues require constant attention and a
concern for their ethical implications. A broad, responsible scientific and social debate needs to take place, one capable of
considering all the available information and of calling things
by their name. It sometimes happens that complete information is not put on the table; a selection is made on the basis
of particular interests, be they politico-economic or ideological. This makes it difficult to reach a balanced and prudent
judgement on different questions, one which takes into account
all the pertinent variables. Discussions are needed in which all
those directly or indirectly affected (farmers, consumers, civil

authorities, scientists, seed producers, people living near fumi-
gated fields, and others) can make known their problems and
concerns, and have access to adequate and reliable information in order to make decisions for the common good, present
and future. This is a complex environmental issue; it calls for
a comprehensive approach which would require, at the very
least, greater efforts to finance various lines of independent, interdisciplinary research capable of shedding new light on the
problem.

136. On the other hand, it is troubling that, when some ecological movements defend the integrity of the environment,
rightly demanding that certain limits be imposed on scientific
research, they sometimes fail to apply those same principles
to human life. There is a tendency to justify transgressing all
boundaries when experimentation is carried out on living human embryos. We forget that the inalienable worth of a human
being transcends his or her degree of development. In the same
way, when technology disregards the great ethical principles, it
ends up considering any practice whatsoever as licit. As we have
seen in this chapter, a technology severed from ethics will not
easily be able to limit its own power.

CHAPTER FOUR

Integral Ecology

137. Since everything is closely interrelated, and today's problems call for a vision capable of taking into account every aspect of the global crisis, I suggest that we now consider some elements of an integral ecology, one which clearly respects its human and social dimensions.

I. Environmental, Economic and Social Ecology

138. Ecology studies the relationship between living organisms and the environment in which they develop. This necessarily entails reflection and debate about the conditions required for the life and survival of society, and the honesty needed to question certain models of development, production and consumption. It cannot be emphasized enough how everything is intercon-

nected. Time and space are not independent of one another, and not even atoms or subatomic particles can be considered in isolation. Just as the different aspects of the planet—physical, chemical and biological—are interrelated, so too living species are part of a network which we will never fully explore and understand. A good part of our genetic code is shared by many living beings. It follows that the fragmentation of knowledge and the isolation of bits of information can actually become a form of ignorance, unless they are integrated into a broader vision of reality.

139. When we speak of the "environment," what we really mean is a relationship existing between nature and the society which lives in it. Nature cannot be regarded as something separate from ourselves or as a mere setting in which we live. We are part of nature, included in it and thus in constant interaction with it. Recognizing the reasons why a given area is polluted requires a study of the workings of society, its economy, its behaviour patterns, and the ways it grasps reality. Given the scale of change, it is no longer possible to find a specific, discrete answer for each part of the problem. It is essential to seek comprehensive solutions which consider the interactions within natural systems themselves and with social systems. We are faced not with two separate crises, one environmental and the other social, but rather with one complex crisis which is both social and environmental. Strategies for a solution demand an integrated approach to combating poverty, restoring dignity to the excluded, and at the same time protecting nature.

140. Due to the number and variety of factors to be taken into account when determining the environmental impact of a con-

crete undertaking, it is essential to give researchers their due role, to facilitate their interaction, and to ensure broad academic freedom. Ongoing research should also give us a better understanding of how different creatures relate to one another in making up the larger units which today we term "ecosystems." We take these systems into account not only to determine how best to use them, but also because they have an intrinsic value independent of their usefulness. Each organism, as a creature of God, is good and admirable in itself; the same is true of the harmonious ensemble of organisms existing in a defined space and functioning as a system. Although we are often not aware of it, we depend on these larger systems for our own existence. We need only recall how ecosystems interact in dispersing carbon dioxide, purifying water, controlling illnesses and epidemics, forming soil, breaking down waste, and in many other ways which we overlook or simply do not know about. Once they become conscious of this, many people realize that we live and act on the basis of a reality which has previously been given to us, which precedes our existence and our abilities. So, when we speak of "sustainable use," consideration must always be given to each ecosystem's regenerative ability in its different areas and aspects.

141. Economic growth, for its part, tends to produce predictable reactions and a certain standardization with the aim of simplifying procedures and reducing costs. This suggests the need for an "economic ecology" capable of appealing to a broader vision of reality. The protection of the environment is in fact "an integral part of the development process and cannot be considered in isolation from it."[114] We urgently need a humanism capable of

bringing together the different fields of knowledge, including
economics, in the service of a more integral and integrating vision. Today, the analysis of environmental problems cannot be
separated from the analysis of human, family, work-related and
urban contexts, nor from how individuals relate to themselves,
which leads in turn to how they relate to others and to the
environment. There is an interrelation between ecosystems and
between the various spheres of social interaction, demonstrat-
ing yet again that "the whole is greater than the part."[115]

142. If everything is related, then the health of a society's institutions has consequences for the environment and the quality
of human life. "Every violation of solidarity and civic friendship harms the environment."[116] In this sense, social ecology
is necessarily institutional, and gradually extends to the whole
of society, from the primary social group, the family, to the
wider local, national and international communities. Within
each social stratum, and between them, institutions develop to
regulate human relationships. Anything which weakens those
institutions has negative consequences, such as injustice, violence and loss of freedom. A number of countries have a relatively low level of institutional effectiveness, which results in
greater problems for their people while benefiting those who
profit from this situation. Whether in the administration of
the state, the various levels of civil society, or relationships between individuals themselves, lack of respect for the law is becoming more common. Laws may be well framed yet remain
a dead letter. Can we hope, then, that in such cases, legislation and regulations dealing with the environment will really
prove effective? We know, for example, that countries which

have clear legislation about the protection of forests continue to keep silent as they watch laws repeatedly being broken. Moreover, what takes place in any one area can have a direct or indirect influence on other areas. Thus, for example, drug use in affluent societies creates a continual and growing demand for products imported from poorer regions, where behaviour is corrupted, lives are destroyed, and the environment continues to deteriorate.

II. Cultural Ecology

143. Together with the patrimony of nature, there is also an historic, artistic and cultural patrimony which is likewise under threat. This patrimony is a part of the shared identity of each place and a foundation upon which to build a habitable city. It is not a matter of tearing down and building new cities, supposedly more respectful of the environment yet not always more attractive to live in. Rather, there is a need to incorporate the history, culture and architecture of each place, thus preserving its original identity. Ecology, then, also involves protecting the cultural treasures of humanity in the broadest sense. More specifically, it calls for greater attention to local cultures when studying environmental problems, favouring a dialogue between scientific-technical language and the language of the people. Culture is more than what we have inherited from the past; it is also, and above all, a living, dynamic and participatory present reality, which cannot be excluded as we rethink the relationship between human beings and the environment.

144. A consumerist vision of human beings, encouraged by the mechanisms of today's globalized economy, has a levelling effect on cultures, diminishing the immense variety which is the heritage of all humanity. Attempts to resolve all problems through uniform regulations or technical interventions can lead to overlooking the complexities of local problems which demand the active participation of all members of the community. New processes taking shape cannot always fit into frameworks imported from outside; they need to be based in the local culture itself. As life and the world are dynamic realities, so our care for the world must also be flexible and dynamic. Merely technical solutions run the risk of addressing symptoms and not the more serious underlying problems. There is a need to respect the rights of peoples and cultures, and to appreciate that the development of a social group presupposes an historical process which takes place within a cultural context and demands the constant and active involvement of local people *from within their proper culture*. Nor can the notion of the quality of life be imposed from without, for quality of life must be understood within the world of symbols and customs proper to each human group.

145. Many intensive forms of environmental exploitation and degradation not only exhaust the resources which provide local communities with their livelihood, but also undo the social structures which, for a long time, shaped cultural identity and their sense of the meaning of life and community. The disappearance of a culture can be just as serious, or even more serious, than the disappearance of a species of plant or animal. The imposition of a dominant lifestyle linked to a single form of production can be just as harmful as the altering of ecosystems.

146. In this sense, it is essential to show special care for indigenous communities and their cultural traditions. They are not merely one minority among others, but should be the principal dialogue partners, especially when large projects affecting their land are proposed. For them, land is not a commodity but rather a gift from God and from their ancestors who rest there, a sacred space with which they need to interact if they are to maintain their identity and values. When they remain on their land, they themselves care for it best. Nevertheless, in various parts of the world, pressure is being put on them to abandon their homelands to make room for agricultural or mining projects which are undertaken without regard for the degradation of nature and culture.

III. Ecology of Daily Life

147. Authentic development includes efforts to bring about an integral improvement in the quality of human life, and this entails considering the setting in which people live their lives. These settings influence the way we think, feel and act. In our rooms, our homes, our workplaces and neighbourhoods, we use our environment as a way of expressing our identity. We make every effort to adapt to our environment, but when it is disorderly, chaotic or saturated with noise and ugliness, such overstimulation makes it difficult to find ourselves integrated and happy.

148. An admirable creativity and generosity is shown by persons and groups who respond to environmental limitations by

alleviating the adverse effects of their surroundings and learning to orient their lives amid disorder and uncertainty. For example, in some places, where the façades of buildings are derelict, people show great care for the interior of their homes, or find contentment in the kindness and friendliness of others. A wholesome social life can light up a seemingly undesirable environment. At times a commendable human ecology is practised by the poor despite numerous hardships. The feeling of asphyxiation brought on by densely populated residential areas is countered if close and warm relationships develop, if communities are created, if the limitations of the environment are compensated for in the interior of each person who feels held within a network of solidarity and belonging. In this way, any place can turn from being a hell on earth into the setting for a dignified life.

149. The extreme poverty experienced in areas lacking harmony, open spaces or potential for integration, can lead to incidents of brutality and to exploitation by criminal organizations. In the unstable neighbourhoods of mega-cities, the daily experience of overcrowding and social anonymity can create a sense of uprootedness which spawns antisocial behaviour and violence. Nonetheless, I wish to insist that love always proves more powerful. Many people in these conditions are able to weave bonds of belonging and togetherness which convert overcrowding into an experience of community in which the walls of the ego are torn down and the barriers of selfishness overcome. This experience of a communitarian salvation often generates creative ideas for the improvement of a building or a neighbourhood.[117]

150. Given the interrelationship between living space and human behaviour, those who design buildings, neighbourhoods, public spaces and cities, ought to draw on the various disciplines which help us to understand people's thought processes, symbolic language and ways of acting. It is not enough to seek the beauty of design. More precious still is the service we offer to another kind of beauty: people's quality of life, their adaptation to the environment, encounter and mutual assistance. Here, too, we see how important it is that urban planning always take into consideration the views of those who will live in these areas.

151. There is also a need to protect those common areas, visual landmarks and urban landscapes which increase our sense of belonging, of rootedness, of "feeling at home" within a city which includes us and brings us together. It is important that the different parts of a city be well integrated and that those who live there have a sense of the whole, rather than being confined to one neighbourhood and failing to see the larger city as space which they share with others. Interventions which affect the urban or rural landscape should take into account how various elements combine to form a whole which is perceived by its inhabitants as a coherent and meaningful framework for their lives. Others will then no longer be seen as strangers, but as part of a "we" which all of us are working to create. For this same reason, in both urban and rural settings, it is helpful to set aside some places which can be preserved and protected from constant changes brought by human intervention.

152. Lack of housing is a grave problem in many parts of the world, both in rural areas and in large cities, since state budgets

usually cover only a small portion of the demand. Not only the poor, but many other members of society as well, find it difficult to own a home. Having a home has much to do with a sense of personal dignity and the growth of families. This is a major issue for human ecology. In some places, where makeshift shanty towns have sprung up, this will mean developing those neighbourhoods rather than razing or displacing them. When the poor live in unsanitary slums or in dangerous tenements, "in cases where it is necessary to relocate them, in order not to heap suffering upon suffering, adequate information needs to be given beforehand, with choices of decent housing offered, and the people directly involved must be part of the process."[118] At the same time, creativity should be shown in integrating rundown neighbourhoods into a welcoming city: "How beautiful those cities which overcome paralyzing mistrust, integrate those who are different and make this very integration a new factor of development! How attractive are those cities which, even in their architectural design, are full of spaces which connect, relate and favour the recognition of others!"[119]

153. The quality of life in cities has much to do with systems of transport, which are often a source of much suffering for those who use them. Many cars, used by one or more people, circulate in cities, causing traffic congestion, raising the level of pollution, and consuming enormous quantities of non-renewable energy. This makes it necessary to build more roads and parking areas which spoil the urban landscape. Many specialists agree on the need to give priority to public transportation. Yet some measures needed will not prove easily acceptable to

society unless substantial improvements are made in the systems themselves, which in many cities force people to put up with undignified conditions due to crowding, inconvenience, infrequent service and lack of safety.

154. Respect for our dignity as human beings often jars with the chaotic realities that people have to endure in city life. Yet this should not make us overlook the abandonment and neglect also experienced by some rural populations which lack access to essential services and where some workers are reduced to conditions of servitude, without rights or even the hope of a more dignified life.

155. Human ecology also implies another profound reality: the relationship between human life and the moral law, which is inscribed in our nature and is necessary for the creation of a more dignified environment. Pope Benedict XVI spoke of an "ecology of man," based on the fact that "man too has a nature that he must respect and that he cannot manipulate at will."[120] It is enough to recognize that our body itself establishes us in a direct relationship with the environment and with other living beings. The acceptance of our bodies as God's gift is vital for welcoming and accepting the entire world as a gift from the Father and our common home, whereas thinking that we enjoy absolute power over our own bodies turns, often subtly, into thinking that we enjoy absolute power over creation. Learning to accept our body, to care for it and to respect its fullest meaning, is an essential element of any genuine human ecology. Also, valuing one's own body in its femininity or masculinity is necessary if I am going to be able to recognize

myself in an encounter with someone who is different. In this way we can joyfully accept the specific gifts of another man or woman, the work of God the Creator, and find mutual enrichment. It is not a healthy attitude which would seek "to cancel out sexual difference because it no longer knows how to confront it."[121]

IV. The Principle of the Common Good

156. An integral ecology is inseparable from the notion of the common good, a central and unifying principle of social ethics. The common good is "the sum of those conditions of social life which allow social groups and their individual members relatively thorough and ready access to their own fulfilment."[122]

157. Underlying the principle of the common good is respect for the human person as such, endowed with basic and inalienable rights ordered to his or her integral development. It has also to do with the overall welfare of society and the development of a variety of intermediate groups, applying the principle of subsidiarity. Outstanding among those groups is the family, as the basic cell of society. Finally, the common good calls for social peace, the stability and security provided by a certain order which cannot be achieved without particular concern for distributive justice; whenever this is violated, violence always ensues. Society as a whole, and the state in particular, are obliged to defend and promote the common good.

158. In the present condition of global society, where injustices abound and growing numbers of people are deprived of basic human rights and considered expendable, the principle of the common good immediately becomes, logically and inevitably, a summons to solidarity and a preferential option for the poorest of our brothers and sisters. This option entails recognizing the implications of the universal destination of the world's goods, but, as I mentioned in the Apostolic Exhortation *Evangelii Gaudium*,[123] it demands before all else an appreciation of the immense dignity of the poor in the light of our deepest convictions as believers. We need only look around us to see that, today, this option is in fact an ethical imperative essential for effectively attaining the common good.

V. Justice Between the Generations

159. The notion of the common good also extends to future generations. The global economic crises have made painfully obvious the detrimental effects of disregarding our common destiny, which cannot exclude those who come after us. We can no longer speak of sustainable development apart from intergenerational solidarity. Once we start to think about the kind of world we are leaving to future generations, we look at things differently; we realize that the world is a gift which we have freely received and must share with others. Since the world has been given to us, we can no longer view reality in a purely utilitarian way, in which efficiency and productivity are entirely

geared to our individual benefit. Intergenerational solidarity is not optional, but rather a basic question of justice, since the world we have received also belongs to those who will follow us. The Portuguese bishops have called upon us to acknowledge this obligation of justice: "The environment is part of a logic of receptivity. It is on loan to each generation, which must then hand it on to the next."[124] An integral ecology is marked by this broader vision.

160. What kind of world do we want to leave to those who come after us, to children who are now growing up? This question not only concerns the environment in isolation; the issue cannot be approached piecemeal. When we ask ourselves what kind of world we want to leave behind, we think in the first place of its general direction, its meaning and its values. Unless we struggle with these deeper issues, I do not believe that our concern for ecology will produce significant results. But if these issues are courageously faced, we are led inexorably to ask other pointed questions: What is the purpose of our life in this world? Why are we here? What is the goal of our work and all our efforts? What need does the earth have of us? It is no longer enough, then, simply to state that we should be concerned for future generations. We need to see that what is at stake is our own dignity. Leaving an inhabitable planet to future generations is, first and foremost, up to us. The issue is one which dramatically affects us, for it has to do with the ultimate meaning of our earthly sojourn.

161. Doomsday predictions can no longer be met with irony or disdain. We may well be leaving to coming generations debris,

desolation and filth. The pace of consumption, waste and environmental change has so stretched the planet's capacity that our contemporary lifestyle, unsustainable as it is, can only precipitate catastrophes, such as those which even now periodically occur in different areas of the world. The effects of the present imbalance can only be reduced by our decisive action, here and now. We need to reflect on our accountability before those who will have to endure the dire consequences.

162. Our difficulty in taking up this challenge seriously has much to do with an ethical and cultural decline which has accompanied the deterioration of the environment. Men and women of our postmodern world run the risk of rampant individualism, and many problems of society are connected with today's self-centred culture of instant gratification. We see this in the crisis of family and social ties and the difficulties of recognizing the other. Parents can be prone to impulsive and wasteful consumption, which then affects their children who find it increasingly difficult to acquire a home of their own and build a family. Furthermore, our inability to think seriously about future generations is linked to our inability to broaden the scope of our present interests and to give consideration to those who remain excluded from development. Let us not only keep the poor of the future in mind, but also today's poor, whose life on this earth is brief and who cannot keep on waiting. Hence, "in addition to a fairer sense of intergenerational solidarity there is also an urgent moral need for a renewed sense of intragenerational solidarity."[125]

CHAPTER FIVE

Lines of Approach and Action

163. So far I have attempted to take stock of our present situation, pointing to the cracks in the planet that we inhabit as well as to the profoundly human causes of environmental degradation. Although the contemplation of this reality in itself has already shown the need for a change of direction and other courses of action, now we shall try to outline the major paths of dialogue which can help us escape the spiral of self-destruction which currently engulfs us.

I. Dialogue on the Environment in the International Community

164. Beginning in the middle of the last century and overcoming many difficulties, there has been a growing conviction that

our planet is a homeland and that humanity is one people living in a common home. An interdependent world not only makes us more conscious of the negative effects of certain lifestyles and models of production and consumption which affect us all; more importantly, it motivates us to ensure that solutions are proposed from a global perspective, and not simply to defend the interests of a few countries. Interdependence obliges us to think of *one world with a common plan*. Yet the same ingenuity which has brought about enormous technological progress has so far proved incapable of finding effective ways of dealing with grave environmental and social problems worldwide. A global consensus is essential for confronting the deeper problems, which cannot be resolved by unilateral actions on the part of individual countries. Such a consensus could lead, for example, to planning a sustainable and diversified agriculture, developing renewable and less polluting forms of energy, encouraging a more efficient use of energy, promoting a better management of marine and forest resources, and ensuring universal access to drinking water.

165. We know that technology based on the use of highly polluting fossil fuels—especially coal, but also oil and, to a lesser degree, gas—needs to be progressively replaced without delay. Until greater progress is made in developing widely accessible sources of renewable energy, it is legitimate to choose the less harmful alternative or to find short-term solutions. But the international community has still not reached adequate agreements about the responsibility for paying the costs of this energy transition. In recent decades, environmental issues have given rise to considerable public debate and have elicited

a variety of committed and generous civic responses. Politics and business have been slow to react in a way commensurate with the urgency of the challenges facing our world. Although the post-industrial period may well be remembered as one of the most irresponsible in history, nonetheless there is reason to hope that humanity at the dawn of the twenty-first century will be remembered for having generously shouldered its grave responsibilities.

166. Worldwide, the ecological movement has made significant advances, thanks also to the efforts of many organizations of civil society. It is impossible here to mention them all, or to review the history of their contributions. But thanks to their efforts, environmental questions have increasingly found a place on public agendas and encouraged more far-sighted approaches. This notwithstanding, recent World Summits on the environment have not lived up to expectations because, due to lack of political will, they were unable to reach truly meaningful and effective global agreements on the environment.

167. The 1992 Earth Summit in Rio de Janeiro is worth mentioning. It proclaimed that "human beings are at the centre of concerns for sustainable development."[126] Echoing the 1972 Stockholm Declaration, it enshrined international cooperation to care for the ecosystem of the entire earth, the obligation of those who cause pollution to assume its costs, and the duty to assess the environmental impact of given projects and works. It set the goal of limiting greenhouse gas concentration in the atmosphere, in an effort to reverse the trend of global warming. It also drew up an agenda with an action plan and a conven-

tion on biodiversity, and stated principles regarding forests. Although the summit was a real step forward, and prophetic for its time, its accords have been poorly implemented, due to the lack of suitable mechanisms for oversight, periodic review and penalties in cases of non-compliance. The principles which it proclaimed still await an efficient and flexible means of practical implementation.

168. Among positive experiences in this regard, we might mention, for example, the Basel Convention on hazardous wastes, with its system of reporting, standards and controls. There is also the binding Convention on international trade in endangered species of wild fauna and flora, which includes on-site visits for verifying effective compliance. Thanks to the Vienna Convention for the protection of the ozone layer and its implementation through the Montreal Protocol and amendments, the problem of the layer's thinning seems to have entered a phase of resolution.

169. As far as the protection of biodiversity and issues related to desertification are concerned, progress has been far less significant. With regard to climate change, the advances have been regrettably few. Reducing greenhouse gases requires honesty, courage and responsibility, above all on the part of those countries which are more powerful and pollute the most. The Conference of the United Nations on Sustainable Development, "Rio+20" (Rio de Janeiro 2012), issued a wide-ranging but ineffectual outcome document. International negotiations cannot make significant progress due to positions taken by countries which place their national interests above the global

common good. Those who will have to suffer the consequences of what we are trying to hide will not forget this failure of conscience and responsibility. Even as this Encyclical was being prepared, the debate was intensifying. We believers cannot fail to ask God for a positive outcome to the present discussions, so that future generations will not have to suffer the effects of our ill-advised delays.

170. Some strategies for lowering pollutant gas emissions call for the internationalization of environmental costs, which would risk imposing on countries with fewer resources burdensome commitments to reducing emissions comparable to those of the more industrialized countries. Imposing such measures penalizes those countries most in need of development. A further injustice is perpetrated under the guise of protecting the environment. Here also, the poor end up paying the price. Furthermore, since the effects of climate change will be felt for a long time to come, even if stringent measures are taken now, some countries with scarce resources will require assistance in adapting to the effects already being produced, which affect their economies. In this context, there is a need for common and differentiated responsibilities. As the bishops of Bolivia have stated, "the countries which have benefited from a high degree of industrialization, at the cost of enormous emissions of greenhouse gases, have a greater responsibility for providing a solution to the problems they have caused."[127]

171. The strategy of buying and selling "carbon credits" can lead to a new form of speculation which would not help reduce the emission of polluting gases worldwide. This system seems to

provide a quick and easy solution under the guise of a certain commitment to the environment, but in no way does it allow for the radical change which present circumstances require. Rather, it may simply become a ploy which permits maintaining the excessive consumption of some countries and sectors.

172. For poor countries, the priorities must be to eliminate extreme poverty and to promote the social development of their people. At the same time, they need to acknowledge the scandalous level of consumption in some privileged sectors of their population and to combat corruption more effectively. They are likewise bound to develop less polluting forms of energy production, but to do so they require the help of countries which have experienced great growth at the cost of the ongoing pollution of the planet. Taking advantage of abundant solar energy will require the establishment of mechanisms and subsidies which allow developing countries access to technology transfer, technical assistance and financial resources, but in a way which respects their concrete situations, since "the compatibility of [infrastructures] with the context for which they have been designed is not always adequately assessed."[128] The costs of this would be low, compared to the risks of climate change. In any event, these are primarily ethical decisions, rooted in solidarity between all peoples.

173. Enforceable international agreements are urgently needed, since local authorities are not always capable of effective intervention. Relations between states must be respectful of each other's sovereignty, but must also lay down mutually agreed means of averting regional disasters which would eventually

affect everyone. Global regulatory norms are needed to impose obligations and prevent unacceptable actions, for example, when powerful companies or countries dump contaminated waste or offshore polluting industries in other countries.

174. Let us also mention the system of governance of the oceans. International and regional conventions do exist, but fragmentation and the lack of strict mechanisms of regulation, control and penalization end up undermining these efforts. The growing problem of marine waste and the protection of the open seas represent particular challenges. What is needed, in effect, is an agreement on systems of governance for the whole range of so-called "global commons."

175. The same mindset which stands in the way of making radical decisions to reverse the trend of global warming also stands in the way of achieving the goal of eliminating poverty. A more responsible overall approach is needed to deal with both problems: the reduction of pollution and the development of poorer countries and regions. The twenty-first century, while maintaining systems of governance inherited from the past, is witnessing a weakening of the power of nation states, chiefly because the economic and financial sectors, being transnational, tends to prevail over the political. Given this situation, it is essential to devise stronger and more efficiently organized international institutions, with functionaries who are appointed fairly by agreement among national governments, and empowered to impose sanctions. As Benedict XVI has affirmed in continuity with the social teaching of the Church: "To manage the global economy; to revive economies hit by the crisis; to

avoid any deterioration of the present crisis and the greater imbalances that would result; to bring about integral and timely
disarmament, food security and peace; to guarantee the protection of the environment and to regulate migration: for all
this, there is urgent need of a true world political authority,
as my predecessor Blessed John XXIII indicated some years
ago."[129] Diplomacy also takes on new importance in the work of
developing international strategies which can anticipate serious
problems affecting us all.

II. Dialogue for New National and Local Policies

176. There are not just winners and losers among countries, but
within poorer countries themselves. Hence different responsibilities need to be identified. Questions related to the environment and economic development can no longer be approached
only from the standpoint of differences between countries; they
also call for greater attention to policies on the national and
local levels.

177. Given the real potential for a misuse of human abilities,
individual states can no longer ignore their responsibility for
planning, coordination, oversight and enforcement within their
respective borders. How can a society plan and protect its future amid constantly developing technological innovations?
One authoritative source of oversight and coordination is the
law, which lays down rules for admissible conduct in the light

of the common good. The limits which a healthy, mature and sovereign society must impose are those related to foresight and security, regulatory norms, timely enforcement, the elimination of corruption, effective responses to undesired side effects of production processes, and appropriate intervention where potential or uncertain risks are involved. There is a growing jurisprudence dealing with the reduction of pollution by business activities. But political and institutional frameworks do not exist simply to avoid bad practice, but also to promote best practice, to stimulate creativity in seeking new solutions and to encourage individual or group initiatives.

178. A politics concerned with immediate results, supported by consumerist sectors of the population, is driven to produce short-term growth. In response to electoral interests, governments are reluctant to upset the public with measures which could affect the level of consumption or create risks for foreign investment. The myopia of power politics delays the inclusion of a far-sighted environmental agenda within the overall agenda of governments. Thus we forget that "time is greater than space,"[130] that we are always more effective when we generate processes rather than holding on to positions of power. True statecraft is manifest when, in difficult times, we uphold high principles and think of the long-term common good. Political powers do not find it easy to assume this duty in the work of nation-building.

179. In some places, cooperatives are being developed to exploit renewable sources of energy which ensure local self-sufficiency and even the sale of surplus energy. This simple example shows

that, while the existing world order proves powerless to assume
its responsibilities, local individuals and groups can make a real
difference. They are able to instil a greater sense of responsibility, a strong sense of community, a readiness to protect others, a spirit of creativity and a deep love for the land. They are
also concerned about what they will eventually leave to their
children and grandchildren. These values are deeply rooted
in indigenous peoples. Because the enforcement of laws is at
times inadequate due to corruption, public pressure has to be
exerted in order to bring about decisive political action. Society, through non-governmental organizations and intermediate groups, must put pressure on governments to develop more
rigorous regulations, procedures and controls. Unless citizens
control political power—national, regional and municipal—it
will not be possible to control damage to the environment. Local legislation can be more effective, too, if agreements exist
between neighbouring communities to support the same environmental policies.

180. There are no uniform recipes, because each country or region has its own problems and limitations. It is also true that
political realism may call for transitional measures and technologies, so long as these are accompanied by the gradual framing
and acceptance of binding commitments. At the same time,
on the national and local levels, much still needs to be done,
such as promoting ways of conserving energy. These would include favouring forms of industrial production with maximum
energy efficiency and diminished use of raw materials, remov-
ing from the market products which are less energy efficient or
more polluting, improving transport systems, and encouraging

the construction and repair of buildings aimed at reducing their energy consumption and levels of pollution. Political activity on the local level could also be directed to modifying consumption, developing an economy of waste disposal and recycling, protecting certain species and planning a diversified agriculture and the rotation of crops. Agriculture in poorer regions can be improved through investment in rural infrastructures, a better organization of local or national markets, systems of irrigation, and the development of techniques of sustainable agriculture. New forms of cooperation and community organization can be encouraged in order to defend the interests of small producers and preserve local ecosystems from destruction. Truly, much can be done!

181. Here, continuity is essential, because policies related to climate change and environmental protection cannot be altered with every change of government. Results take time and demand immediate outlays which may not produce tangible effects within any one government's term. That is why, in the absence of pressure from the public and from civic institutions, political authorities will always be reluctant to intervene, all the more when urgent needs must be met. To take up these responsibilities and the costs they entail, politicians will inevitably clash with the mindset of short-term gain and results which dominates present-day economics and politics. But if they are courageous, they will attest to their God-given dignity and leave behind a testimony of selfless responsibility. A healthy politics is sorely needed, capable of reforming and coordinating institutions, promoting best practices and overcoming undue pressure and bureaucratic inertia. It should be added, though,

that even the best mechanisms can break down when there are no worthy goals and values, or a genuine and profound humanism to serve as the basis of a noble and generous society.

III. Dialogue and Transparency in Decision-Making

182. An assessment of the environmental impact of business ventures and projects demands transparent political processes involving a free exchange of views. On the other hand, the forms of corruption which conceal the actual environmental impact of a given project, in exchange for favours, usually produce specious agreements which fail to inform adequately and to allow for full debate.

183. Environmental impact assessment should not come after the drawing up of a business proposition or the proposal of a particular policy, plan or programme. It should be part of the process from the beginning, and be carried out in a way which is interdisciplinary, transparent and free of all economic or political pressure. It should be linked to a study of working conditions and possible effects on people's physical and mental health, on the local economy and on public safety. Economic returns can thus be forecast more realistically, taking into account potential scenarios and the eventual need for further investment to correct possible undesired effects. A consensus should always be reached between the different stakeholders, who can offer a variety of approaches, solutions and alternatives. The local

population should have a special place at the table; they are concerned about their own future and that of their children, and can consider goals transcending immediate economic interest. We need to stop thinking in terms of "interventions" to save the environment in favour of policies developed and debated by all interested parties. The participation of the latter also entails being fully informed about such projects and their different risks and possibilities; this includes not just preliminary decisions but also various follow-up activities and continued monitoring. Honesty and truth are needed in scientific and political discussions; these should not be limited to the issue of whether or not a particular project is permitted by law.

184. In the face of possible risks to the environment which may affect the common good now and in the future, decisions must be made "based on a comparison of the risks and benefits foreseen for the various possible alternatives."[131] This is especially the case when a project may lead to a greater use of natural resources, higher levels of emission or discharge, an increase of refuse, or significant changes to the landscape, the habitats of protected species or public spaces. Some projects, if insufficiently studied, can profoundly affect the quality of life of an area due to very different factors such as unforeseen noise pollution, the shrinking of visual horizons, the loss of cultural values, or the effects of nuclear energy use. The culture of consumerism, which prioritizes short-term gain and private interest, can make it easy to rubber-stamp authorizations or to conceal information.

185. In any discussion about a proposed venture, a number of questions need to be asked in order to discern whether or not

it will contribute to genuine integral development. What will it accomplish? Why? Where? When? How? For whom? What are the risks? What are the costs? Who will pay those costs and how? In this discernment, some questions must have higher priority. For example, we know that water is a scarce and indispensable resource and a fundamental right which conditions the exercise of other human rights. This indisputable fact overrides any other assessment of environmental impact on a region.

186. The Rio Declaration of 1992 states that "where there are threats of serious or irreversible damage, lack of full scientific certainty shall not be used as a pretext for postponing cost-effective measures"[132] which prevent environmental degradation. This precautionary principle makes it possible to protect those who are most vulnerable and whose ability to defend their interests and to assemble incontrovertible evidence is limited. If objective information suggests that serious and irreversible damage may result, a project should be halted or modified, even in the absence of indisputable proof. Here the burden of proof is effectively reversed, since in such cases objective and conclusive demonstrations will have to be brought forward to demonstrate that the proposed activity will not cause serious harm to the environment or to those who inhabit it.

187. This does not mean being opposed to any technological innovations which can bring about an improvement in the quality of life. But it does mean that profit cannot be the sole criterion to be taken into account, and that, when significant new information comes to light, a reassessment should be made, with

the involvement of all interested parties. The outcome may be a decision not to proceed with a given project, to modify it or to consider alternative proposals.

188. There are certain environmental issues where it is not easy to achieve a broad consensus. Here I would state once more that the Church does not presume to settle scientific questions or to replace politics. But I am concerned to encourage an honest and open debate so that particular interests or ideologies will not prejudice the common good.

IV. Politics and Economy in Dialogue for Human Fulfilment

189. Politics must not be subject to the economy, nor should the economy be subject to the dictates of an efficiency-driven paradigm of technocracy. Today, in view of the common good, there is urgent need for politics and economics to enter into a frank dialogue in the service of life, especially human life. Saving banks at any cost, making the public pay the price, foregoing a firm commitment to reviewing and reforming the entire system, only reaffirms the absolute power of a financial system, a power which has no future and will only give rise to new crises after a slow, costly and only apparent recovery. The financial crisis of 2007–08 provided an opportunity to develop a new economy, more attentive to ethical principles, and new ways of regulating speculative financial practices and virtual wealth.

But the response to the crisis did not include rethinking the outdated criteria which continue to rule the world. Production is not always rational, and is usually tied to economic variables which assign to products a value that does not necessarily correspond to their real worth. This frequently leads to an overproduction of some commodities, with unnecessary impact on the environment and with negative results on regional economies.[133] The financial bubble also tends to be a productive bubble. The problem of the real economy is not confronted with vigour, yet it is the real economy which makes diversification and improvement in production possible, helps companies to function well, and enables small and medium businesses to develop and create employment.

190. Here too, it should always be kept in mind that "environmental protection cannot be assured solely on the basis of financial calculations of costs and benefits. The environment is one of those goods that cannot be adequately safeguarded or promoted by market forces."[134] Once more, we need to reject a magical conception of the market, which would suggest that problems can be solved simply by an increase in the profits of companies or individuals. Is it realistic to hope that those who are obsessed with maximizing profits will stop to reflect on the environmental damage which they will leave behind for future generations? Where profits alone count, there can be no thinking about the rhythms of nature, its phases of decay and regeneration, or the complexity of ecosystems which may be gravely upset by human intervention. Moreover, biodiversity is considered at most a deposit of economic resources available for exploitation, with no serious thought for the real value of

things, their significance for persons and cultures, or the concerns and needs of the poor.

191. Whenever these questions are raised, some react by accusing others of irrationally attempting to stand in the way of progress and human development. But we need to grow in the conviction that a decrease in the pace of production and consumption can at times give rise to another form of progress and development. Efforts to promote a sustainable use of natural resources are not a waste of money, but rather an investment capable of providing other economic benefits in the medium term. If we look at the larger picture, we can see that more diversified and innovative forms of production which impact less on the environment can prove very profitable. It is a matter of openness to different possibilities which do not involve stifling human creativity and its ideals of progress, but rather directing that energy along new channels.

192. For example, a path of productive development, which is more creative and better directed, could correct the present disparity between excessive technological investment in consumption and insufficient investment in resolving urgent problems facing the human family. It could generate intelligent and profitable ways of reusing, revamping and recycling, and it could also improve the energy efficiency of cities. Productive diversification offers the fullest possibilities to human ingenuity to create and innovate, while at the same time protecting the environment and creating more sources of employment. Such creativity would be a worthy expression of our most noble human qualities, for we would be striving intelligently, boldly and

responsibly to promote a sustainable and equitable development within the context of a broader concept of quality of life. On the other hand, to find ever new ways of despoiling nature, purely for the sake of new consumer items and quick profit, would be, in human terms, less worthy and creative, and more superficial.

193. In any event, if in some cases sustainable development were to involve new forms of growth, then in other cases, given the insatiable and irresponsible growth produced over many decades, we need also to think of containing growth by setting some reasonable limits and even retracing our steps before it is too late. We know how unsustainable is the behaviour of those who constantly consume and destroy, while others are not yet able to live in a way worthy of their human dignity. That is why the time has come to accept decreased growth in some parts of the world, in order to provide resources for other places to experience healthy growth. Benedict XVI has said that "technologically advanced societies must be prepared to encourage more sober lifestyles, while reducing their energy consumption and improving its efficiency."[135]

194. For new models of progress to arise, there is a need to change "models of global development"[136]; this will entail a responsible reflection on "the meaning of the economy and its goals with an eye to correcting its malfunctions and misapplications."[137] It is not enough to balance, in the medium term, the protection of nature with financial gain, or the preservation of the environment with progress. Halfway measures simply delay the inevitable disaster. Put simply, it is a matter of redefining our

notion of progress. A technological and economic development which does not leave in its wake a better world and an integrally higher quality of life cannot be considered progress. Frequently, in fact, people's quality of life actually diminishes—by the deterioration of the environment, the low quality of food or the depletion of resources—in the midst of economic growth. In this context, talk of sustainable growth usually becomes a way of distracting attention and offering excuses. It absorbs the language and values of ecology into the categories of finance and technocracy, and the social and environmental responsibility of businesses often gets reduced to a series of marketing and image-enhancing measures.

195. The principle of the maximization of profits, frequently isolated from other considerations, reflects a misunderstanding of the very concept of the economy. As long as production is increased, little concern is given to whether it is at the cost of future resources or the health of the environment; as long as the clearing of a forest increases production, no one calculates the losses entailed in the desertification of the land, the harm done to biodiversity or the increased pollution. In a word, businesses profit by calculating and paying only a fraction of the costs involved. Yet only when "the economic and social costs of using up shared environmental resources are recognized with transparency and fully borne by those who incur them, not by other peoples or future generations,"[138] can those actions be considered ethical. An instrumental way of reasoning, which provides a purely static analysis of realities in the service of present needs, is at work whether resources are allocated by the market or by state central planning.

196. What happens with politics? Let us keep in mind the principle of subsidiarity, which grants freedom to develop the capabilities present at every level of society, while also demanding a greater sense of responsibility for the common good from those who wield greater power. Today, it is the case that some economic sectors exercise more power than states themselves. But economics without politics cannot be justified, since this would make it impossible to favour other ways of handling the various aspects of the present crisis. The mindset which leaves no room for sincere concern for the environment is the same mindset which lacks concern for the inclusion of the most vulnerable members of society. For "the current model, with its emphasis on success and self-reliance, does not appear to favour an investment in efforts to help the slow, the weak or the less talented to find opportunities in life."[139]

197. What is needed is a politics which is far-sighted and capable of a new, integral and interdisciplinary approach to handling the different aspects of the crisis. Often, politics itself is responsible for the disrepute in which it is held, on account of corruption and the failure to enact sound public policies. If in a given region the state does not carry out its responsibilities, some business groups can come forward in the guise of benefactors, wield real power, and consider themselves exempt from certain rules, to the point of tolerating different forms of organized crime, human trafficking, the drug trade and violence, all of which become very difficult to eradicate. If politics shows itself incapable of breaking such a perverse logic, and remains caught up in inconsequential discussions, we will continue to avoid facing the major problems of humanity. A strategy for real

change calls for rethinking processes in their entirety, for it is not enough to include a few superficial ecological considerations while failing to question the logic which underlies present-day culture. A healthy politics needs to be able to take up this challenge.

198. Politics and the economy tend to blame each other when it comes to poverty and environmental degradation. It is to be hoped that they can acknowledge their own mistakes and find forms of interaction directed to the common good. While some are concerned only with financial gain, and others with holding on to or increasing their power, what we are left with are conflicts or spurious agreements where the last thing either party is concerned about is caring for the environment and protecting those who are most vulnerable. Here too, we see how true it is that "unity is greater than conflict."[140]

V. Religions in Dialogue with Science

199. It cannot be maintained that empirical science provides a complete explanation of life, the interplay of all creatures and the whole of reality. This would be to breach the limits imposed by its own methodology. If we reason only within the confines of the latter, little room would be left for aesthetic sensibility, poetry, or even reason's ability to grasp the ultimate meaning and purpose of things.[141] I would add that "religious classics can prove meaningful in every age; they have an enduring power

to open new horizons . . . Is it reasonable and enlightened to dismiss certain writings simply because they arose in the context of religious belief?"[142] It would be quite simplistic to think that ethical principles present themselves purely in the abstract, detached from any context. Nor does the fact that they may be couched in religious language detract from their value in public debate. The ethical principles capable of being apprehended by reason can always reappear in different guise and find expression in a variety of languages, including religious language.

200. Any technical solution which science claims to offer will be powerless to solve the serious problems of our world if humanity loses its compass, if we lose sight of the great motivations which make it possible for us to live in harmony, to make sacrifices and to treat others well. Believers themselves must constantly feel challenged to live in a way consonant with their faith and not to contradict it by their actions. They need to be encouraged to be ever open to God's grace and to draw constantly from their deepest convictions about love, justice and peace. If a mistaken understanding of our own principles has at times led us to justify mistreating nature, to exercise tyranny over creation, to engage in war, injustice and acts of violence, we believers should acknowledge that by so doing we were not faithful to the treasures of wisdom which we have been called to protect and preserve. Cultural limitations in different eras often affected the perception of these ethical and spiritual treasures, yet by constantly returning to their sources, religions will be better equipped to respond to today's needs.

201. The majority of people living on our planet profess to be believers. This should spur religions to dialogue among themselves for the sake of protecting nature, defending the poor, and building networks of respect and fraternity. Dialogue among the various sciences is likewise needed, since each can tend to become enclosed in its own language, while specialization leads to a certain isolation and the absolutization of its own field of knowledge. This prevents us from confronting environmental problems effectively. An open and respectful dialogue is also needed between the various ecological movements, among which ideological conflicts are not infrequently encountered. The gravity of the ecological crisis demands that we all look to the common good, embarking on a path of dialogue which demands patience, self-discipline and generosity, always keeping in mind that "realities are greater than ideas."[143]

CHAPTER SIX

Ecological Education and Spirituality

202. Many things have to change course, but it is we human beings above all who need to change. We lack an awareness of our common origin, of our mutual belonging, and of a future to be shared with everyone. This basic awareness would enable the development of new convictions, attitudes and forms of life. A great cultural, spiritual and educational challenge stands before us, and it will demand that we set out on the long path of renewal.

I. Towards a New Lifestyle

203. Since the market tends to promote extreme consumerism in an effort to sell its products, people can easily get caught up in a whirlwind of needless buying and spending. Compulsive

consumerism is one example of how the techno-economic paradigm affects individuals. Romano Guardini had already foreseen this: "The gadgets and technics forced upon him by the patterns of machine production and of abstract planning mass man accepts quite simply; they are the forms of life itself. To either a greater or lesser degree mass man is convinced that his conformity is both reasonable and just."[144] This paradigm leads people to believe that they are free as long as they have the supposed freedom to consume. But those really free are the minority who wield economic and financial power. Amid this confusion, postmodern humanity has not yet achieved a new self-awareness capable of offering guidance and direction, and this lack of identity is a source of anxiety. We have too many means and only a few insubstantial ends.

204. The current global situation engenders a feeling of instability and uncertainty, which in turn becomes "a seedbed for collective selfishness."[145] When people become self-centred and self-enclosed, their greed increases. The emptier a person's heart is, the more he or she needs things to buy, own and consume. It becomes almost impossible to accept the limits imposed by reality. In this horizon, a genuine sense of the common good also disappears. As these attitudes become more widespread, social norms are respected only to the extent that they do not clash with personal needs. So our concern cannot be limited merely to the threat of extreme weather events, but must also extend to the catastrophic consequences of social unrest. Obsession with a consumerist lifestyle, above all when few people are capable of maintaining it, can only lead to violence and mutual destruction.

205. Yet all is not lost. Human beings, while capable of the worst, are also capable of rising above themselves, choosing again what is good, and making a new start, despite their mental and social conditioning. We are able to take an honest look at ourselves, to acknowledge our deep dissatisfaction, and to embark on new paths to authentic freedom. No system can completely suppress our openness to what is good, true and beautiful, or our God-given ability to respond to his grace at work deep in our hearts. I appeal to everyone throughout the world not to forget this dignity which is ours. No one has the right to take it from us.

206. A change in lifestyle could bring healthy pressure to bear on those who wield political, economic and social power. This is what consumer movements accomplish by boycotting certain products. They prove successful in changing the way businesses operate, forcing them to consider their environmental footprint and their patterns of production. When social pressure affects their earnings, businesses clearly have to find ways to produce differently. This shows us the great need for a sense of social responsibility on the part of consumers. "Purchasing is always a moral—and not simply economic—act."[146] Today, in a word, "the issue of environmental degradation challenges us to examine our lifestyle."[147]

207. The Earth Charter asked us to leave behind a period of self-destruction and make a new start, but we have not as yet developed a universal awareness needed to achieve this. Here, I would echo that courageous challenge: "As never before in history, common destiny beckons us to seek a new beginning . . .

Let ours be a time remembered for the awakening of a new
reverence for life, the firm resolve to achieve sustainability, the
quickening of the struggle for justice and peace, and the joyful
celebration of life."[148]

208. We are always capable of going out of ourselves towards
the other. Unless we do this, other creatures will not be recognized for their true worth; we are unconcerned about caring for
things for the sake of others; we fail to set limits on ourselves in
order to avoid the suffering of others or the deterioration of our
surroundings. Disinterested concern for others, and the rejec-
tion of every form of self-centeredness and self-absorption, are
essential if we truly wish to care for our brothers and sisters and
for the natural environment. These attitudes also attune us to
the moral imperative of assessing the impact of our every action
and personal decision on the world around us. If we can overcome individualism, we will truly be able to develop a different
lifestyle and bring about significant changes in society.

II. Educating for the Covenant Between Humanity and the Environment

209. An awareness of the gravity of today's cultural and ecological crisis must be translated into new habits. Many people
know that our current progress and the mere amassing of things
and pleasures are not enough to give meaning and joy to the
human heart, yet they feel unable to give up what the market

sets before them. In those countries which should be making the greatest changes in consumer habits, young people have a new ecological sensitivity and a generous spirit, and some of them are making admirable efforts to protect the environment. At the same time, they have grown up in a milieu of extreme consumerism and affluence which makes it difficult to develop other habits. We are faced with an educational challenge.

210. Environmental education has broadened its goals. Whereas in the beginning it was mainly centred on scientific information, consciousness-raising and the prevention of environmental risks, it tends now to include a critique of the "myths" of a modernity grounded in a utilitarian mindset (individualism, unlimited progress, competition, consumerism, the unregulated market). It seeks also to restore the various levels of ecological equilibrium, establishing harmony within ourselves, with others, with nature and other living creatures, and with God. Environmental education should facilitate making the leap towards the transcendent which gives ecological ethics its deepest meaning. It needs educators capable of developing an ethics of ecology, and helping people, through effective pedagogy, to grow in solidarity, responsibility and compassionate care.

211. Yet this education, aimed at creating an "ecological citizenship," is at times limited to providing information, and fails to instil good habits. The existence of laws and regulations is insufficient in the long run to curb bad conduct, even when effective means of enforcement are present. If the laws are to bring about significant, long-lasting effects, the majority of the

members of society must be adequately motivated to accept them, and personally transformed to respond. Only by cultivating sound virtues will people be able to make a selfless ecological commitment. A person who could afford to spend and consume more but regularly uses less heating and wears warmer clothes, shows the kind of convictions and attitudes which help to protect the environment. There is a nobility in the duty to care for creation through little daily actions, and it is wonderful how education can bring about real changes in lifestyle. Education in environmental responsibility can encourage ways of acting which directly and significantly affect the world around us, such as avoiding the use of plastic and paper, reducing water consumption, separating refuse, cooking only what can reasonably be consumed, showing care for other living beings, using public transport or car-pooling, planting trees, turning off unnecessary lights, or any number of other practices. All of these reflect a generous and worthy creativity which brings out the best in human beings. Reusing something instead of immediately discarding it, when done for the right reasons, can be an act of love which expresses our own dignity.

212. We must not think that these efforts are not going to change the world. They benefit society, often unbeknown to us, for they call forth a goodness which, albeit unseen, inevitably tends to spread. Furthermore, such actions can restore our sense of self-esteem; they can enable us to live more fully and to feel that life on earth is worthwhile.

213. Ecological education can take place in a variety of settings: at school, in families, in the media, in catechesis and elsewhere.

Good education plants seeds when we are young, and these continue to bear fruit throughout life. Here, though, I would stress the great importance of the family, which is "the place in which life—the gift of God—can be properly welcomed and protected against the many attacks to which it is exposed, and can develop in accordance with what constitutes authentic human growth. In the face of the so-called culture of death, the family is the heart of the culture of life."[149] In the family we first learn how to show love and respect for life; we are taught the proper use of things, order and cleanliness, respect for the local ecosystem and care for all creatures. In the family we receive an integral education, which enables us to grow harmoniously in personal maturity. In the family we learn to ask without demanding, to say "thank you" as an expression of genuine gratitude for what we have been given, to control our aggressivity and greed, and to ask forgiveness when we have caused harm. These simple gestures of heartfelt courtesy help to create a culture of shared life and respect for our surroundings.

214. Political institutions and various other social groups are also entrusted with helping to raise people's awareness. So too is the Church. All Christian communities have an important role to play in ecological education. It is my hope that our seminaries and houses of formation will provide an education in responsible simplicity of life, in grateful contemplation of God's world, and in concern for the needs of the poor and the protection of the environment. Because the stakes are so high, we need institutions empowered to impose penalties for damage inflicted on the environment. But we also need the personal qualities of self-control and willingness to learn from one another.

215. In this regard, "the relationship between a good aesthetic education and the maintenance of a healthy environment cannot be overlooked."[150] By learning to see and appreciate beauty, we learn to reject self-interested pragmatism. If someone has not learned to stop and admire something beautiful, we should not be surprised if he or she treats everything as an object to be used and abused without scruple. If we want to bring about deep change, we need to realize that certain mindsets really do influence our behaviour. Our efforts at education will be inadequate and ineffectual unless we strive to promote a new way of thinking about human beings, life, society and our relationship with nature. Otherwise, the paradigm of consumerism will continue to advance, with the help of the media and the highly effective workings of the market.

III. Ecological Conversion

216. The rich heritage of Christian spirituality, the fruit of twenty centuries of personal and communal experience, has a precious contribution to make to the renewal of humanity. Here, I would like to offer Christians a few suggestions for an ecological spirituality grounded in the convictions of our faith, since the teachings of the Gospel have direct consequences for our way of thinking, feeling and living. More than in ideas or concepts as such, I am interested in how such a spirituality can motivate us to a more passionate concern for the protection of our world. A commitment this lofty cannot be sustained by doctrine alone, without a spirituality capable of inspiring

us, without an "interior impulse which encourages, motivates, nourishes and gives meaning to our individual and communal activity."[151] Admittedly, Christians have not always appropriated and developed the spiritual treasures bestowed by God upon the Church, where the life of the spirit is not dissociated from the body or from nature or from worldly realities, but lived in and with them, in communion with all that surrounds us.

217. "The external deserts in the world are growing, because the internal deserts have become so vast."[152] For this reason, the ecological crisis is also a summons to profound interior conversion. It must be said that some committed and prayerful Christians, with the excuse of realism and pragmatism, tend to ridicule expressions of concern for the environment. Others are passive; they choose not to change their habits and thus become inconsistent. So what they all need is an "ecological conversion," whereby the effects of their encounter with Jesus Christ become evident in their relationship with the world around them. Living our vocation to be protectors of God's handiwork is essential to a life of virtue; it is not an optional or a secondary aspect of our Christian experience.

218. In calling to mind the figure of Saint Francis of Assisi, we come to realize that a healthy relationship with creation is one dimension of overall personal conversion, which entails the recognition of our errors, sins, faults and failures, and leads to heartfelt repentance and desire to change. The Australian bishops spoke of the importance of such conversion for achieving reconciliation with creation: "To achieve such reconciliation, we must examine our lives and acknowledge the ways in which

we have harmed God's creation through our actions and our
failure to act. We need to experience a conversion, or change
of heart."[153]

219. Nevertheless, self-improvement on the part of individuals will not by itself remedy the extremely complex situation
facing our world today. Isolated individuals can lose their ability and freedom to escape the utilitarian mindset, and end up
prey to an unethical consumerism bereft of social or ecological
awareness. Social problems must be addressed by community
networks and not simply by the sum of individual good deeds.
This task "will make such tremendous demands of man that
he could never achieve it by individual initiative or even by the
united effort of men bred in an individualistic way. The work
of dominating the world calls for a union of skills and a unity
of achievement that can only grow from quite a different attitude."[154] The ecological conversion needed to bring about lasting
change is also a community conversion.

220. This conversion calls for a number of attitudes which together foster a spirit of generous care, full of tenderness. First,
it entails gratitude and gratuitousness, a recognition that the
world is God's loving gift, and that we are called quietly to
imitate his generosity in self-sacrifice and good works: "Do not
let your left hand know what your right hand is doing . . . and
your Father who sees in secret will reward you" (Mt 6:3–4). It
also entails a loving awareness that we are not disconnected
from the rest of creatures, but joined in a splendid universal
communion. As believers, we do not look at the world from
without but from within, conscious of the bonds with which

the Father has linked us to all beings. By developing our individual, God-given capacities, an ecological conversion can
inspire us to greater creativity and enthusiasm in resolving the
world's problems and in offering ourselves to God "as a living
sacrifice, holy and acceptable" (*Rom* 12:1). We do not understand our superiority as a reason for personal glory or irresponsible dominion, but rather as a different capacity which,
in its turn, entails a serious responsibility stemming from our
faith.

221. Various convictions of our faith, developed at the beginning of this Encyclical can help us to enrich the meaning of
this conversion. These include the awareness that each creature
reflects something of God and has a message to convey to us,
and the security that Christ has taken unto himself this material world and now, risen, is intimately present to each being,
surrounding it with his affection and penetrating it with his
light. Then too, there is the recognition that God created the
world, writing into it an order and a dynamism that human
beings have no right to ignore. We read in the Gospel that
Jesus says of the birds of the air that "not one of them is forgotten before God" (*Lk* 12:6). How then can we possibly mistreat
them or cause them harm? I ask all Christians to recognize
and to live fully this dimension of their conversion. May the
power and the light of the grace we have received also be evident in our relationship to other creatures and to the world
around us. In this way, we will help nurture that sublime fraternity with all creation which Saint Francis of Assisi so radi-
antly embodied.

IV. Joy and Peace

222. Christian spirituality proposes an alternative understanding of the quality of life, and encourages a prophetic and contemplative lifestyle, one capable of deep enjoyment free of the obsession with consumption. We need to take up an ancient lesson, found in different religious traditions and also in the Bible. It is the conviction that "less is more." A constant flood of new consumer goods can baffle the heart and prevent us from cherishing each thing and each moment. To be serenely present to each reality, however small it may be, opens us to much greater horizons of understanding and personal fulfilment. Christian spirituality proposes a growth marked by moderation and the capacity to be happy with little. It is a return to that simplicity which allows us to stop and appreciate the small things, to be grateful for the opportunities which life affords us, to be spiritually detached from what we possess, and not to succumb to sadness for what we lack. This implies avoiding the dynamic of dominion and the mere accumulation of pleasures.

223. Such sobriety, when lived freely and consciously, is liberating. It is not a lesser life or one lived with less intensity. On the contrary, it is a way of living life to the full. In reality, those who enjoy more and live better each moment are those who have given up dipping here and there, always on the look-out for what they do not have. They experience what it means to appreciate each person and each thing, learning familiarity with the simplest things and how to enjoy them. So they are able to shed unsatisfied needs, reducing their obsessiveness and weariness. Even living on little, they can live a lot, above all

when they cultivate other pleasures and find satisfaction in fraternal encounters, in service, in developing their gifts, in music and art, in contact with nature, in prayer. Happiness means knowing how to limit some needs which only diminish us, and being open to the many different possibilities which life can offer.

224. Sobriety and humility were not favourably regarded in the last century. And yet, when there is a general breakdown in the exercise of a certain virtue in personal and social life, it ends up causing a number of imbalances, including environmental ones. That is why it is no longer enough to speak only of the integrity of ecosystems. We have to dare to speak of the integrity of human life, of the need to promote and unify all the great values. Once we lose our humility, and become enthralled with the possibility of limitless mastery over everything, we inevitably end up harming society and the environment. It is not easy to promote this kind of healthy humility or happy sobriety when we consider ourselves autonomous, when we exclude God from our lives or replace him with our own ego, and think that our subjective feelings can define what is right and what is wrong.

225. On the other hand, no one can cultivate a sober and satisfying life without being at peace with him or herself. An adequate understanding of spirituality consists in filling out what we mean by peace, which is much more than the absence of war. Inner peace is closely related to care for ecology and for the common good because, lived out authentically, it is reflected in a balanced lifestyle together with a capacity for wonder which takes us to a deeper understanding of life. Nature is

filled with words of love, but how can we listen to them amid constant noise, interminable and nerve-wracking distractions, or the cult of appearances? Many people today sense a profound imbalance which drives them to frenetic activity and makes them feel busy, in a constant hurry which in turn leads them to ride rough-shod over everything around them. This too affects how they treat the environment. An integral ecology includes taking time to recover a serene harmony with creation, reflecting on our lifestyle and our ideals, and contemplating the Creator who lives among us and surrounds us, whose presence "must not be contrived but found, uncovered."[155]

226. We are speaking of an attitude of the heart, one which approaches life with serene attentiveness, which is capable of being fully present to someone without thinking of what comes next, which accepts each moment as a gift from God to be lived to the full. Jesus taught us this attitude when he invited us to contemplate the lilies of the field and the birds of the air, or when seeing the rich young man and knowing his restlessness, "he looked at him with love" (*Mk* 10:21). He was completely present to everyone and to everything, and in this way he showed us the way to overcome that unhealthy anxiety which makes us superficial, aggressive and compulsive consumers.

227. One expression of this attitude is when we stop and give thanks to God before and after meals. I ask all believers to return to this beautiful and meaningful custom. That moment of blessing, however brief, reminds us of our dependence on God for life; it strengthens our feeling of gratitude for the gifts of creation; it acknowledges those who by their labours provide us

with these goods; and it reaffirms our solidarity with those in greatest need.

V. Civic and Political Love

228. Care for nature is part of a lifestyle which includes the capacity for living together and communion. Jesus reminded us that we have God as our common Father and that this makes us brothers and sisters. Fraternal love can only be gratuitous; it can never be a means of repaying others for what they have done or will do for us. That is why it is possible to love our enemies. This same gratuitousness inspires us to love and accept the wind, the sun and the clouds, even though we cannot control them. In this sense, we can speak of a "universal fraternity."

229. We must regain the conviction that we need one another, that we have a shared responsibility for others and the world, and that being good and decent are worth it. We have had enough of immorality and the mockery of ethics, goodness, faith and honesty. It is time to acknowledge that light-hearted superficiality has done us no good. When the foundations of social life are corroded, what ensues are battles over conflicting interests, new forms of violence and brutality, and obstacles to the growth of a genuine culture of care for the environment.

230. Saint Therese of Lisieux invites us to practise the little way of love, not to miss out on a kind word, a smile or any small gesture which sows peace and friendship. An integral ecology

is also made up of simple daily gestures which break with the logic of violence, exploitation and selfishness. In the end, a world of exacerbated consumption is at the same time a world which mistreats life in all its forms.

231. Love, overflowing with small gestures of mutual care, is also civic and political, and it makes itself felt in every action that seeks to build a better world. Love for society and commitment to the common good are outstanding expressions of a charity which affects not only relationships between individuals but also "macro-relationships, social, economic and political ones."[156] That is why the Church set before the world the ideal of a "civilization of love."[157] Social love is the key to authentic development: "In order to make society more human, more worthy of the human person, love in social life—political, economic and cultural—must be given renewed value, becoming the constant and highest norm for all activity."[158] In this framework, along with the importance of little everyday gestures, social love moves us to devise larger strategies to halt environmental degradation and to encourage a "culture of care" which permeates all of society. When we feel that God is calling us to intervene with others in these social dynamics, we should realize that this too is part of our spirituality, which is an exercise of charity and, as such, matures and sanctifies us.

232. Not everyone is called to engage directly in political life. Society is also enriched by a countless array of organizations which work to promote the common good and to defend the environment, whether natural or urban. Some, for example, show concern for a public place (a building, a fountain, an aban-

doned monument, a landscape, a square), and strive to protect, restore, improve or beautify it as something belonging to everyone. Around these community actions, relationships develop or are recovered and a new social fabric emerges. Thus, a community can break out of the indifference induced by consumerism. These actions cultivate a shared identity, with a story which can be remembered and handed on. In this way, the world, and the quality of life of the poorest, are cared for, with a sense of solidarity which is at the same time aware that we live in a common home which God has entrusted to us. These community actions, when they express self-giving love, can also become intense spiritual experiences.

VI. Sacramental Signs and the Celebration of Rest

233. The universe unfolds in God, who fills it completely. Hence, there is a mystical meaning to be found in a leaf, in a mountain trail, in a dewdrop, in a poor person's face.[159] The ideal is not only to pass from the exterior to the interior to discover the action of God in the soul, but also to discover God in all things. Saint Bonaventure teaches us that "contemplation deepens the more we feel the working of God's grace within our hearts, and the better we learn to encounter God in creatures outside ourselves."[160]

234. Saint John of the Cross taught that all the goodness present in the realities and experiences of this world "is present in

God eminently and infinitely, or more properly, in each of these sublime realities is God."[161] This is not because the finite things of this world are really divine, but because the mystic experiences the intimate connection between God and all beings, and thus feels that "all things are God."[162] Standing awestruck before a mountain, he or she cannot separate this experience from God, and perceives that the interior awe being lived has to be entrusted to the Lord: "Mountains have heights and they are plentiful, vast, beautiful, graceful, bright and fragrant. These mountains are what my Beloved is to me. Lonely valleys are quiet, pleasant, cool, shady and flowing with fresh water; in the variety of their groves and in the sweet song of the birds, they afford abundant recreation and delight to the senses, and in their solitude and silence, they refresh us and give rest. These valleys are what my Beloved is to me."[163]

235. The Sacraments are a privileged way in which nature is taken up by God to become a means of mediating supernatural life. Through our worship of God, we are invited to embrace the world on a different plane. Water, oil, fire and colours are taken up in all their symbolic power and incorporated in our act of praise. The hand that blesses is an instrument of God's love and a reflection of the closeness of Jesus Christ, who came to accompany us on the journey of life. Water poured over the body of a child in Baptism is a sign of new life. Encountering God does not mean fleeing from this world or turning our back on nature. This is especially clear in the spirituality of the Christian East. "Beauty, which in the East is one of the best loved names expressing the divine harmony and the model of humanity transfigured, appears everywhere: in the shape of

a church, in the sounds, in the colours, in the lights, in the scents."[164] For Christians, all the creatures of the material universe find their true meaning in the incarnate Word, for the Son of God has incorporated in his person part of the material world, planting in it a seed of definitive transformation. "Christianity does not reject matter. Rather, bodiliness is considered in all its value in the liturgical act, whereby the human body is disclosed in its inner nature as a temple of the Holy Spirit and is united with the Lord Jesus, who himself took a body for the world's salvation."[165]

236. It is in the Eucharist that all that has been created finds its greatest exaltation. Grace, which tends to manifest itself tangibly, found unsurpassable expression when God himself became man and gave himself as food for his creatures. The Lord, in the culmination of the mystery of the Incarnation, chose to reach our intimate depths through a fragment of matter. He comes not from above, but from within, he comes that we might find him in this world of ours. In the Eucharist, fullness is already achieved; it is the living centre of the universe, the overflowing core of love and of inexhaustible life. Joined to the incarnate Son, present in the Eucharist, the whole cosmos gives thanks to God. Indeed the Eucharist is itself an act of cosmic love: "Yes, cosmic! Because even when it is celebrated on the humble altar of a country church, the Eucharist is always in some way celebrated on the altar of the world."[166] The Eucharist joins heaven and earth; it embraces and penetrates all creation. The world which came forth from God's hands returns to him in blessed and undivided adoration: in the bread of the Eucharist, "creation is projected towards divinization, towards the holy

wedding feast, towards unification with the Creator himself."[167]
Thus, the Eucharist is also a source of light and motivation for
our concerns for the environment, directing us to be stewards
of all creation.

237. On Sunday, our participation in the Eucharist has special
importance. Sunday, like the Jewish Sabbath, is meant to be a
day which heals our relationships with God, with ourselves,
with others and with the world. Sunday is the day of the Res-
urrection, the "first day" of the new creation, whose first fruits
are the Lord's risen humanity, the pledge of the final transfigu-
ration of all created reality. It also proclaims "man's eternal rest
in God."[168] In this way, Christian spirituality incorporates the
value of relaxation and festivity. We tend to demean contemplative rest as something unproductive and unnecessary, but
this is to do away with the very thing which is most important
about work: its meaning. We are called to include in our work
a dimension of receptivity and gratuity, which is quite different from mere inactivity. Rather, it is another way of working,
which forms part of our very essence. It protects human action
from becoming empty activism; it also prevents that unfettered
greed and sense of isolation which make us seek personal gain
to the detriment of all else. The law of weekly rest forbade work
on the seventh day, "so that your ox and your donkey may have
rest, and the son of your maidservant, and the stranger, may be
refreshed" (*Ex* 23:12). Rest opens our eyes to the larger picture
and gives us renewed sensitivity to the rights of others. And so
the day of rest, centred on the Eucharist, sheds it light on the
whole week, and motivates us to greater concern for nature and
the poor.

VII. The Trinity and the Relationship Between Creatures

238. The Father is the ultimate source of everything, the loving and self-communicating foundation of all that exists. The Son, his reflection, through whom all things were created, united himself to this earth when he was formed in the womb of Mary. The Spirit, infinite bond of love, is intimately present at the very heart of the universe, inspiring and bringing new pathways. The world was created by the three Persons acting as a single divine principle, but each one of them performed this common work in accordance with his own personal property. Consequently, "when we contemplate with wonder the universe in all its grandeur and beauty, we must praise the whole Trinity."[169]

239. For Christians, believing in one God who is trinitarian communion suggests that the Trinity has left its mark on all creation. Saint Bonaventure went so far as to say that human beings, before sin, were able to see how each creature "testifies that God is three." The reflection of the Trinity was there to be recognized in nature "when that book was open to man and our eyes had not yet become darkened."[170] The Franciscan saint teaches us that *each creature bears in itself a specifically Trinitarian structure*, so real that it could be readily contemplated if only the human gaze were not so partial, dark and fragile. In this way, he points out to us the challenge of trying to read reality in a Trinitarian key.

240. The divine Persons are subsistent relations, and the world, created according to the divine model, is a web of relationships. Creatures tend towards God, and in turn it is proper to every living being to tend towards other things, so that throughout the universe we can find any number of constant and secretly interwoven relationships.[171] This leads us not only to marvel at the manifold connections existing among creatures, but also to discover a key to our own fulfilment. The human person grows more, matures more and is sanctified more to the extent that he or she enters into relationships, going out from themselves to live in communion with God, with others and with all creatures. In this way, they make their own that trinitarian dynamism which God imprinted in them when they were created. Everything is interconnected, and this invites us to develop a spirituality of that global solidarity which flows from the mystery of the Trinity.

VIII. Queen of All Creation

241. Mary, the Mother who cared for Jesus, now cares with maternal affection and pain for this wounded world. Just as her pierced heart mourned the death of Jesus, so now she grieves for the sufferings of the crucified poor and for the creatures of this world laid waste by human power. Completely transfigured, she now lives with Jesus, and all creatures sing of her fairness. She is the Woman, "clothed in the sun, with the moon under her feet, and on her head a crown of twelve stars" (*Rev* 12:1). Carried up into heaven, she is the Mother and Queen

of all creation. In her glorified body, together with the Risen Christ, part of creation has reached the fullness of its beauty. She treasures the entire life of Jesus in her heart (cf. *Lk* 2:19, 51), and now understands the meaning of all things. Hence, we can ask her to enable us to look at this world with eyes of wisdom.

242. At her side in the Holy Family of Nazareth, stands the figure of Saint Joseph. Through his work and generous presence, he cared for and defended Mary and Jesus, delivering them from the violence of the unjust by bringing them to Egypt. The Gospel presents Joseph as a just man, hard-working and strong. But he also shows great tenderness, which is not a mark of the weak but of those who are genuinely strong, fully aware of reality and ready to love and serve in humility. That is why he was proclaimed custodian of the universal Church. He too can teach us how to show care; he can inspire us to work with generosity and tenderness in protecting this world which God has entrusted to us.

IX. Beyond the Sun

243. At the end, we will find ourselves face to face with the infinite beauty of God (cf. *1 Cor* 13:12), and be able to read with admiration and happiness the mystery of the universe, which with us will share in unending plenitude. Even now we are journeying towards the sabbath of eternity, the new Jerusalem, towards our common home in heaven. Jesus says: "I make all things new" (*Rev* 21:5). Eternal life will be a shared experience

of awe, in which each creature, resplendently transfigured, will take its rightful place and have something to give those poor men and women who will have been liberated once and for all.

244. In the meantime, we come together to take charge of this home which has been entrusted to us, knowing that all the good which exists here will be taken up into the heavenly feast. In union with all creatures, we journey through this land seeking God, for "if the world has a beginning and if it has been created, we must enquire who gave it this beginning, and who was its Creator."[172] Let us sing as we go. May our struggles and our concern for this planet never take away the joy of our hope.

245. God, who calls us to generous commitment and to give him our all, offers us the light and the strength needed to continue on our way. In the heart of this world, the Lord of life, who loves us so much, is always present. He does not abandon us, he does not leave us alone, for he has united himself definitively to our earth, and his love constantly impels us to find new ways forward. *Praise be to him!*

* * * * *

246. At the conclusion of this lengthy reflection which has been both joyful and troubling, I propose that we offer two prayers. The first we can share with all who believe in a God who is the all-powerful Creator, while in the other we Christians ask for inspiration to take up the commitment to creation set before us by the Gospel of Jesus.

A PRAYER FOR OUR EARTH

All-powerful God, you are present in the whole universe
and in the smallest of your creatures.
You embrace with your tenderness all that exists.
Pour out upon us the power of your love,
that we may protect life and beauty.
Fill us with peace, that we may live
as brothers and sisters, harming no one.
O God of the poor,
help us to rescue the abandoned and forgotten of this earth,
so precious in your eyes.
Bring healing to our lives,
that we may protect the world and not prey on it,
that we may sow beauty, not pollution and destruction.
Touch the hearts
of those who look only for gain
at the expense of the poor and the earth.
Teach us to discover the worth of each thing,
to be filled with awe and contemplation,
to recognize that we are profoundly united
with every creature
as we journey towards your infinite light.
We thank you for being with us each day.
Encourage us, we pray, in our struggle
for justice, love and peace.

A CHRISTIAN PRAYER IN UNION WITH CREATION

Father, we praise you with all your creatures.
They came forth from your all-powerful hand;
they are yours, filled with your presence and your tender love.
Praise be to you!

Son of God, Jesus,
through you all things were made.
You were formed in the womb of Mary our Mother,
you became part of this earth,
and you gazed upon this world with human eyes.
Today you are alive in every creature
in your risen glory.
Praise be to you!

Holy Spirit, by your light
you guide this world towards the Father's love
and accompany creation as it groans in travail.
You also dwell in our hearts
and you inspire us to do what is good.
Praise be to you!

Triune Lord, wondrous community of infinite love,
teach us to contemplate you
in the beauty of the universe,
for all things speak of you.
Awaken our praise and thankfulness
for every being that you have made.
Give us the grace to feel profoundly joined
to everything that is.

God of love, show us our place in this world
as channels of your love
for all the creatures of this earth,
for not one of them is forgotten in your sight.
Enlighten those who possess power and money
that they may avoid the sin of indifference,
that they may love the common good, advance the weak,
and care for this world in which we live.
The poor and the earth are crying out.
O Lord, seize us with your power and light,
help us to protect all life,
to prepare for a better future,
for the coming of your Kingdom
of justice, peace, love and beauty.
Praise be to you!
Amen.

Given in Rome at Saint Peter's on 24 May, the Solemnity of Pentecost, in the year 2015, the third of my Pontificate.

Franciscus

Notes

INTRODUCTION

i. Notable exceptions are Bill McKibben, *Deep Economy: The Wealth of Communities and the Durable Future* (St. Martin's Griffin, 2008); James Gustave Speth, *The Bridge at the End of the World: Capitalism, the Environment, and Crossing from Crisis to Sustainability* (Yale University Press, 2009); Paul Gilding, *The Great Disruption: Why the Climate Crisis Will Bring On the End of Shopping and the Birth of a New World* (Bloombury, 2012); and Naomi Klein, *This Changes Everything: Capitalism vs. the Climate* (Simon & Schuster, 2014).

ii. One thinks of Langdon Winner, *The Whale and the Reactor: A Search for Limits in an Age of High Technology* (University of Chicago Press, 1989); Steven Shapin and Simon Schaeffer, *Leviathan and the Air Pump: Hobbes, Boyle and the Experimental Life* (Princeton University Press, 1986).

iii. Francis Fukuyama, *The End of History and the Last Man* (Free Press, 2006; 1992). See also truth-out.org/archive/component/k2/item/81249:the-ideology-of-no-ideology and www

.theguardian.com/books/2014/mar/21/bring-back-ideology-fukuyama-end-history-25-years-on.

iv. data.worldbank.org/sites/default/files/section1.pdf. See also www.theguardian.com/society/2015/jul/01/billions-have-no-access-to-toilets-says-world-health-organisation-report? CMP=twt_gu.

v. For a characteristic example of the equation of abundant energy, consumerism, and freedom used as a marketing tool, see www.energytomorrow.org/energy-101?utm_source=EnergyGuardian_email&utm_medium=email&utm_campaign=13745.

vi. www.amazon.com/Happiness-Guide-Developing-Lifes-Important/dp/0316167258/ref=sr_1_1?ie=UTF8&qid=1435931631&sr=8-1&keywords=happiness.

vii. David Brooks, "Fracking and the Franciscans," June 23, 2015. www.nytimes.com/2015/06/23/opinion/fracking-and-the-franciscans.html?_r=0. See also news.harvard.edu/gazette/story/2015/06/a-blessing-to-slow-climate-change/.

viii. www.nytimes.com/2015/06/24/business/combating-climate-change-with-science-rather-than-hope.html?rref=collection%2Fcolumn%2Feconomic-scene&contentCollection=business&action=click&module=NextInCollection&region=Footer&pgtype=article.

CHAPTER I: WHAT IS HAPPENING TO OUR COMMON HOME

1. *Canticle of the Creatures*, in *Francis of Assisi: Early Documents*, vol. 1, New York–London–Manila, 1999, 113–14.

2. Apostolic Letter *Octogesima Adveniens* (14 May 1971), 21: AAS 63 (1971), 416–17.

3. *Address to FAO on the 25th Anniversary of Its Institution* (16 November 1970), 4: AAS 62 (1970), 833.

4. Encyclical Letter *Redemptor Hominis* (4 March 1979), 15: AAS 71 (1979), 287.

5. Cf. *Catechesis* (17 January 2001), 4: Insegnamenti 41/1 (2001), 179.

6. Encyclical Letter *Centesimus Annus* (1 May 1991), 38: AAS 83 (1991), 841.

7. Ibid., 58: AAS 83 (1991), 863.

8. John Paul II, Encyclical Letter *Sollicitudo Rei Socialis* (30 December 1987), 34: AAS 80 (1988), 559.

9. Cf. id., Encyclical Letter *Centesimus Annus* (1 May 1991), 37: AAS 83 (1991), 840.

10. *Address to the Diplomatic Corps Accredited to the Holy See* (8 January 2007): AAS 99 (2007), 73.

11. Encyclical Letter *Caritas in Veritate* (29 June 2009), 51: AAS 101 (2009), 687.

12. *Address to the Bundestag*, Berlin (22 September 2011): AAS 103 (2011), 664.

13. *Address to the Clergy of the Diocese of Bolzano-Bressanone* (6 August 2008): AAS 100 (2008), 634.

14. *Message for the Day of Prayer for the Protection of Creation* (1 September 2012).

15. *Address in Santa Barbara, California* (8 November 1997); cf. John Chryssavgis, *On Earth as in Heaven: Ecological Vision and Initiatives of Ecumenical Patriarch Bartholomew*, Bronx, New York, 2012.

16. Ibid.

17. *Lecture at the Monastery of Utstein*, Norway (23 June 2003).

18. "Global Responsibility and Ecological Sustainability," Closing Remarks, Halki Summit I, Istanbul (20 June 2012).

19. Thomas of Celano, *The Life of Saint Francis*, I, 29, 81: in *Francis of Assisi: Early Documents*, vol. 1, New York–London–Manila, 1999, 251.

20. *The Major Legend of Saint Francis*, VIII, 6, in *Francis of Assisi: Early Documents*, vol. 2, New York–London–Manila, 2000, 590.

21. Cf. Thomas of Celano, *The Remembrance of the Desire of a Soul*, II, 124, 165, in *Francis of Assisi: Early Documents*, vol. 2, New York–London–Manila, 2000, 354.

22. Southern African Catholic Bishops' Conference, *Pastoral Statement on the Environmental Crisis* (5 September 1999).

23. Cf. *Greeting to the Staff of FAO* (20 November 2014): AAS 106 (2014), 985.

24. Fifth General Conference of the Latin American and Caribbean Bishops, *Aparecida Document* (29 June 2007), 86.

25. Catholic Bishops' Conference of the Philippines, Pastoral Letter *What is Happening to our Beautiful Land?* (29 January 1988).

26. Bolivian Bishops' Conference, Pastoral Letter on the Environment and Human Development in Bolivia *El universo, don de Dios para la vida* (23 March 2012), 17.

27. Cf. German Bishops' Conference, Commission for Social Issues, *Der Klimawandel: Brennpunkt globaler, intergenerationeller und ökologischer Gerechtigkeit* (September 2006), 28–30.

28. Pontifical Council for Justice and Peace, *Compendium of the Social Doctrine of the Church*, 483.

29. *Catechesis* (5 June 2013): *Insegnamenti* 1/1 (2013), 280.

30. Bishops of the Patagonia-Comahue Region (Argentina), *Christmas Message* (December 2009), 2.

31. United States Conference of Catholic Bishops, *Global Climate Change: A Plea for Dialogue, Prudence and the Common Good* (15 June 2001).

32. Fifth General Conference of the Latin American and Caribbean Bishops, *Aparecida Document* (29 June 2007), 471.

33. Apostolic Exhortation *Evangelii Gaudium* (24 November 2013), 56: AAS 105 (2013), 1043.

34. John Paul II, *Message for the 1990 World Day of Peace*, 12: AAS 82 (1990), 154.

35. Id., *Catechesis* (17 January 2001), 3: *Insegnamenti* 24/1 (2001), 178.

CHAPTER 2: THE GOSPEL OF CREATION

36. John Paul II, *Message for the 1990 World Day of Peace*, 15: AAS 82 (1990), 156.

37. *Catechism of the Catholic Church*, 357.
38. *Angelus* in Osnabrück (Germany) with the disabled, 16 November 1980: *Insegnamenti* 3/2 (1980), 1232.
39. Benedict XVI, *Homily for the Solemn Inauguration of the Petrine Ministry* (24 April 2005): AAS 97 (2005), 711.
40. Cf. Bonaventure, *The Major Legend of Saint Francis*, VIII, 1, in *Francis of Assisi: Early Documents*, vol. 2, New York–London–Manila, 2000, 586.
41. *Catechism of the Catholic Church*, 2416.
42. German Bishops' Conference, *Zukunft der Schöpfung—Zukunft der Menschheit. Einklärung der Deutschen Bischofskonferenz zu Fragen der Umwelt und der Energieversorgung* (1980), II, 2.
43. *Catechism of the Catholic Church*, 339.
44. *Hom. in Hexaemeron*, I, 2, 10: PG 29, 9.
45. *The Divine Comedy, Paradiso*, Canto XXXIII, 145.
46. Benedict XVI, *Catechesis* (9 November 2005), 3: *Insegnamenti* 1 (2005), 768.
47. Id., Encyclical Letter *Caritas in Veritate* (29 June 2009), 51: AAS 101 (2009), 687.
48. John Paul II, *Catechesis* (24 April 1991), 6: *Insegnamenti* 14 (1991), 856.
49. The Catechism explains that God wished to create a world which is "journeying towards its ultimate perfection," and that this implies the presence of imperfection and physical evil; cf. *Catechism of the Catholic Church*, 310.
50. Cf. Second Vatican Ecumenical Council, Pastoral Constitution on the Church in the Modern World *Gaudium et Spes*, 36.
51. Thomas Aquinas, *Summa Theologiae*, I, q. 104, art. 1 ad 4.
52. Id., *In octo libros Physicorum Aristotelis expositio*, Lib. II, lectio 14.
53. Against this horizon we can set the contribution of Fr Teilhard de Chardin; cf. Paul VI, *Address in a Chemical and Pharmaceutical Plant* (24 February 1966): *Insegnamenti* 4 (1966), 992–93; John Paul II, *Letter to the Reverend George Coyne* (1 June 1988): *Insegnamenti* 11/2 (1988), 1715; Benedict XVI, *Homily for the*

Celebration of Vespers in Aosta (24 July 2009): *Insegnamenti* 5/2 (2009), 60.

54. John Paul II, *Catechesis* (30 January 2002),6: *Insegnamenti* 25/1 (2002), 140.

55. Canadian Conference of Catholic Bishops, Social Affairs Commission, Pastoral Letter *You Love All That Exists . . . All Things Are Yours, God, Lover of Life* (4 October 2003), 1.

56. Catholic Bishops' Conference of Japan, *Reverence for Life. A Message for the Twenty-First Century* (1 January 2000), 89.

57. John Paul II, *Catechesis* (26 January 2000), 5: *Insegnamenti* 23/1 (2000), 123.

58. Id., *Catechesis* (2 August 2000), 3: *Insegnamenti* 23/2 (2000), 112.

59. Paul Ricoeur, *Philosophie de la Volonté, t. II: Finitude et Culpabilité*, Paris, 2009, 216.

60. *Summa Theologiae*, I, q. 47, art. 1.

61. Ibid.

62. Cf. Ibid., art. 2, ad 1; art. 3.

63. *Catechism of the Catholic Church*, 340.

64. *Canticle of the Creatures*, in *Francis of Assisi: Early Documents*, New York–London–Manila, 1999, 113–14.

65. Cf. National Conference of the Bishops of Brazil, *A Igreja e a Questão Ecológica*, 1992, 53 New York–London–Manila, 54.

66. Ibid., 61.

67. Apostolic Exhortation *Evangelii Gaudium* (24 November 2013), 215: AAS 105 (2013), 1109.

68. Cf. Benedict XVI, Encyclical Letter *Caritas in Veritate* (29 June 2009), 14: AAS 101 (2009), 650.

69. *Catechism of the Catholic Church*, 2418.

70. Conference of Dominican Bishops, Pastoral Letter *Sobre la relación del hombre con la naturaleza* (21 January 1987).

71. John Paul II, Encyclical Letter *Laborem Exercens* (14 September 1981), 19: AAS 73 (1981), 626.

72. Encyclical Letter *Centesimus Annus* (1 May 1991), 31: AAS 83 (1991), 831.

73. Encyclical Letter *Sollicitudo Rei Socialis* (30 December 1987), 33: AAS 80 (1988), 557.

74. *Address to Indigenous and Rural People*, Cuilapán, Mexico (29 January 1979), 6: AAS 71 (1979), 209.

75. *Homily at Mass for Farmers*, Recife, Brazil (7 July 1980): AAS 72 (1980): AAS 72 (1980), 926.

76. Cf. *Message for the 1990 World Day of Peace*, 8: AAS 82 (1990), 152.

77. Paraguayan Bishops' Conference, Pastoral Letter *El campesino paraguayo y la tierra* (12 June 1983), 2, 4, d.

78. New Zealand Catholic Bishops' Conference, *Statement on Environmental Issues* (1 September 2006).

79. Encyclical Letter *Laborem Exercens* (14 September 1981), 27: AAS 73 (1981), 645.

80. Hence Saint Justin could speak of "seeds of the Word" in the world; cf. *II Apologia* 8, 1–2; 13, 3–6: PG 6, 457–58, 467.

CHAPTER 3: THE HUMAN ROOTS OF THE
ECOLOGICAL CRISIS

81. John Paul II, *Address to Scientists and Representatives of the United Nations University*, Hiroshima (25 February 1981), 3: AAS 73 (1981), 422.

82. Benedict XVI, Encyclical Letter *Caritas in Veritate* (29 June 2009), 69: AAS 101 (2009), 702.

83. Romano Guardini, *Das Ende der Neuzeit*, 9th ed., Würzburg, 1965, 87 (English: *The End of the Modern World*, Wilmington, 1998, 82).

84. Ibid.

85. Ibid., 87–88 (*The End of the Modern World*, 83).

86. Pontifical Council for Justice and Peace, *Compendium of the Social Doctrine of the Church*, 462.

87. Romano Guardini, *Das Ende der Neuzeit*, 63–64 (*The End of the Modern World*, 56).

88. Ibid., 64 (*The End of the Modern World*, 56).

89. Cf. Benedict XVI, Encyclical Letter *Caritas in Veritate* (29 June 2009), 35: AAS 101 (2009), 671.

90. Ibid., 22: 657.

91. Apostolic Exhortation *Evangelii Gaudium* (24 November 2013), 231: AAS 105 (2013), 1114.

92. Romano Guardini, *Das Ende der Neuzeit*, 63 (*The End of the Modern World*, 55).

93. John Paul II, Encyclical Letter *Centesimus Annus* (1 May 1991), 38: AAS 83 (1991), 841.

94. Cf. *Love for Creation. An Asian Response to the Ecological Crisis*, Declaration of the Colloquium sponsored by the Federation of Asian Bishops' Conferences (Tagatay, 31 January–5 February 1993), 3.3.2.

95. John Paul II, Encyclical Letter *Centesimus Annus* (1 May 1991), 37: AAS 83 (1991), 840.

96. Benedict XVI, *Message for the 2010 World Day of Peace*, 2: AAS 102 (2010), 41.

97. Id., Encyclical Letter *Caritas in Veritate* (29 June 2009), 28: AAS 101 (2009), 663.

98. Cf. Vincent of Lerins, *Commonitorium Primum*, ch. 23: PL 50, 688: "Ut annis scilicet consolidetur, dilatetur tempore, sublimetur aetate."

99. No. 80: AAS 105 (2013), 1053.

100. Second Vatican Ecumenical Council, Pastoral Constitution on the Church in the Modern World *Gaudium et Spes*, 63.

101. Cf. John Paul II, Encyclical Letter *Centesimus Annus* (1 May 1991), 37: AAS 83 (1991), 840.

102. Paul VI, Encyclical Letter *Populorum Progressio* (26 March 1967), 34: AAS 59 (1967), 274.

103. Benedict XVI, Encyclical Letter *Caritas in Veritate* (29 June 2009), 32: AAS 101 (2009), 666.

104. Ibid.

105. Ibid.

106. *Catechism of the Catholic Church*, 2417.

107. Ibid., 2418.

108. Ibid., 2415.

109. *Message for the 1990 World Day of Peace*, 6: AAS 82 (1990), 150.

110. *Address to the Pontifical Academy of Sciences* (3 October 1981), 3: *Insegnamenti* 4/2 (1981), 333.

111. *Message for the 1990 World Day of Peace*, 7: AAS 82 (1990), 151.

112. John Paul II, *Address to the 35th General Assembly of the World Medical Association* (29 October 1983), 6: AAS 76 (1984), 394.

113. Episcopal Commission for Pastoral Concerns in Argentina, *Una tierra para todos* (June 2005), 19.

CHAPTER 4: INTEGRAL ECOLOGY

114. *Rio Declaration on Environment and Development* (14 June 1992), Principle 4.

115. Apostolic Exhortation *Evangelii Gaudium* (24 November 2013), 237: AAS 105 (2013), 1116.

116. Benedict XVI, Encyclical Letter *Caritas in Veritate* (29 June 2009), 51: AAS 101 (2009), 687.

117. Some authors have emphasized the values frequently found, for example, in the *villas, chabolas* or *favelas* of Latin America: cf. Juan Carlos Scannone, S.J., "La irrupción del pobre y la lógica de la gratuidad," in Juan Carlos Scannone and Marcelo Perine (eds.), *Irrupción del pobre y quehacer filosófico. Hacia una nueva racionalidad*, Buenos Aires, 1993, 225–30.

118. Pontifical Council for Justice and Peace, *Compendium of the Social Doctrine of the Church*, 482.

119. Apostolic Exhortation *Evangelii Gaudium* (24 November 2013), 210: AAS 105 (2013), 1107.

120. *Address to the German Bundestag*, Berlin (22 September 2011): AAS 103 (2011), 668.

121. *Catechesis* (15 April 2015): *L'Osservatore Romano*, 16 April 2015, 8.

122. Second Vatican Ecumenical Council, Pastoral Constitution on the Church in the Modern World *Gaudium et Spes*, 26.

123. Cf. Nos. 186–201: AAS 105 (2013), 1098–1105.

124. Portuguese Bishops' Conference, Pastoral Letter *Responsabili-dade Solidária pelo Bem Comum* (15 September 2003), 20.

125. Benedict XVI, *Message for the 2010 World Day of Peace*, 8: AAS 102 (2010), 45.

CHAPTER 5: LINES OF APPROACH AND ACTION

126. *Rio Declaration on Environment and Development* (14 June 1992), Principle 1.

127. Bolivian Bishops' Conference, Pastoral Letter on the Environment and Human Development in Bolivia *El universo, don de Dios para la vida* (March 2012), 86.

128. Pontifical Council for Justice and Peace, *Energy, Justice and Peace*, IV, 1, Vatican City (2014), 53.

129. Benedict XVI, Encyclical Letter *Caritas in Veritate* (29 June 2009), 67: AAS 101 (2009).

130. Apostolic Exhortation *Evangelii Gaudium* (24 November 2013), 222: AAS 105 (2013), 1111.

131. Pontifical Council for Justice and Peace, *Compendium of the Social Doctrine of the Church*, 469.

132. *Rio Declaration on the Environment and Development* (14 June 1992), Principle 15.

133. Cf. Mexican Bishops' Conferrence, Episcopal Commission for Pastoral and Social Concerns, *Jesucristo, vida y esperanza de los indígenas e campesinos* (14 January 2008).

134. Pontifical Council for Justice and Peace, *Compendium of the Social Doctrine of the Church*, 470.

135. *Message for the 2010 World Day of Peace*, 9: AAS 102 (2010), 46.

136. Ibid.

137. Ibid., 5: 43.

138. Benedict XVI, Encyclical Letter *Caritas in Veritate* (29 June 2009), 50: AAS 101 (2009), 686.

139. Apostolic Exhortation *Evangelii Gaudium* (24 November 2013), 209: AAS 105 (2013), 1107.

140. Ibid., 228: AAS 105 (2013), 1113.

141. Cf. Encyclical Letter *Lumen Fidei* (29 June 2013), 34: AAS 105 (2013), 577: "Nor is the light of faith, joined to the truth of love, extraneous to the material world, for love is always lived out in body and spirit; the light of faith is an incarnate light radiating from the luminous life of Jesus. It also illumines the material world, trusts its inherent order, and knows that it calls us to an ever widening path of harmony and understanding. The gaze of science thus benefits from faith: faith encourages the scientist to remain constantly open to reality in all its inexhaustible richness. Faith awakens the critical sense by preventing research from being satisfied with its own formulae and helps it to realize that nature is always greater. By stimulating wonder before the profound mystery of creation, faith broadens the horizons of reason to shed greater light on the world which discloses itself to scientific investigation."

142. Apostolic Exhortation *Evangelii Gaudium* (24 November 2013), 256: AAS 105 (2013), 1123.

143. Ibid., 231: 1114.

CHAPTER 6: ECOLOGICAL EDUCATION AND
SPIRITUALITY

144. Romano Guardini, *Das Ende der Neuzeit*, 9th edition, Würzburg, 1965, 66–67 (English: *The End of the Modern World*, Wilmington, 1998, 60).

145. John Paul II, *Message for the 1990 World Day of Peace*, 1: AAS 82 (1990), 147.

146. Benedict XVI, Encyclical Letter *Caritas in Veritate* (29 June 2009), 66: AAS 101 (2009), 699.

147. Id., *Message for the 2010 World Day of Peace*, 11: AAS 102 (2010), 48.

148. *Earth Charter*, The Hague (29 June 2000).

149. John Paul II, Encyclical Letter *Centesimus Annus* (1 May 1991), 39: AAS 83 (1991), 842.

150. Id., *Message for the 1990 World Day of Peace*, 14: AAS 82 (1990), 155.

151. Apostolic Exhortation *Evangelii Gaudium* (24 November 2013), 261: AAS 105 (2013), 1124.

152. Benedict XVI, *Homily for the Solemn Inauguration of the Petrine Ministry* (24 April 2005): AAS 97 (2005), 710.

153. Australian Catholic Bishops' Conference, *A New Earth—The Environmental Challenge* (2002).

154. Romano Guardini, *Das Ende der Neuzeit*, 72 (*The End of the Modern World*, 65–66).

155. Apostolic Exhortation *Evangelii Gaudium* (24 November 2013), 71: AAS 105 (2013), 1050.

156. Benedict XVI, Encyclical Letter *Caritas in Veritate* (29 June 2009) 2: AAS 101 (2009), 642.

157. Paul VI, *Message for the 1977 World Day of Peace*: AAS 68 (1976), 709.

158. Pontifical Council for Justice and Peace, *Compendium of the Social Doctrine of the Church*, 582.

159. The spiritual writer Ali al-Khawas stresses from his own experience the need not to put too much distance between the creatures of the world and the interior experience of God. As he puts it: "Prejudice should not have us criticize those who seek ecstasy in music or poetry. There is a subtle mystery in each of the movements and sounds of this world. The initiate will capture what is being said when the wind blows, the trees sway, water flows, flies buzz, doors creak, birds sing, or in the sound of strings or flutes, the sighs of the sick, the groans of the afflicted . . ." (Eva de Vitray-Meyerovitch [ed.], *Anthologie du soufisme*, Paris 1978, 200).

160. *In II Sent.*, 23, 2, 3.

161. *Cántico Espiritual*, XIV, 5.

162. Ibid.

163. Ibid., XIV, 6–7.

164. John Paul II, Apostolic Letter *Orientale Lumen* (2 May 1995), 11: AAS 87 (1995), 757.

165. Ibid.

166. Id., Encyclical Letter *Ecclesia de Eucharistia* (17 April 2003), 8: AAS 95 (2003), 438.

167. Benedict XVI, *Homily for the Mass of Corpus Domini* (15 June 2006): AAS 98 (2006), 513.

168. *Catechism of the Catholic Church*, 2175.

169. John Paul II, *Catechesis* (2 August 2000), 4: *Insegnamenti* 23/2 (2000), 112.

170. *Quaest. Disp. de Myst. Trinitatis*, 1, 2 concl.

171. Cf. Thomas Aquinas, *Summa Theologiae*, I, q. 11, art. 3; q. 21, art. 1, ad 3; q. 47, art. 3.

172. Basil the Great, *Hom. in Hexaemeron*, I, 2, 6: PG 29, 8.

Born Jorge Mario Bergoglio, **POPE FRANCIS** has been the Pope of the Catholic Church since March 13, 2013, when he became the 266th pontiff. He is the first Latin American and the first Jesuit to lead the Roman Catholic Church—and the first non-European leader of the church in 1,200 years. He took the name Francis after St. Francis of Assisi. Born in Buenos Aires in 1936 to Italian immigrant parents, Pope Francis was ordained as a Catholic priest in 1969. He became a bishop in 1992 and the Archbishop of Buenos Aires in 1998, and, in 2001, was appointed as a cardinal by Pope John Paul II. Devoted to rectifying social injustices and economic inequality, Pope Francis has said that he "would like to see a church that is poor and is for the poor."

NAOMI ORESKES is Professor of the History of Science and Affiliated Professor of Earth and Planetary Sciences at Harvard University. Her opinion pieces have been featured in *The New York Times*, *The Washington Post*, *Nature*, *Science*, and other leading publications. She is the author of a number of books, including *Merchants of Doubt: How a Handful of Scientists Obscured the Truth on Issues from Tobacco Smoke to Global Warming*, coauthored with Erik M. Conway. In May 2014, she attended "Sustainable Nature, Sustainable Humanity," a meeting of the Pontifical Academy of Sciences that helped to lay the foundations for Pope Francis's *Encyclical*.